Printing with Power:
The Future of
Metal Manufacturing

Namita

Table of Contents

Chapter 1.   Introduction 1

Chapter 2.   Background 4

2.1   Additive Manufacturing Processes 4

   2.1.1   DED Processing Using Wire Feed 6

   2.1.2   GTAW AM DED Processing 10

2.2   Numerical Modelling of Heat Profile 12

2.3   Heat Treatment and Microstructure 16

   2.3.3   Equilibrium Conditions 16

   2.3.4   Non-Equilibrium Conditions 18

   2.3.5   Effects of Thermal Cycling 23

Chapter 3.   Experimental Methodology 25

3.1   Numerical Modeling 25

3.2   Material 28

3.3   Weld Process Control Methodology 29

3.4   GTAW Based MAM Platform 31

3.5   Control System 33

3.6   Data Acquisition System 34

3.7   Data Processing 37

3.8   Test Specimen Nomenclature 38

3.9   Hardness Testing 39

3.10   Microscopy 42

3.11   Test Matrix 43

Chapter 4.   Experimental Results 45

4.1   Modeled and As-Built Thermal Profiles 45

4.1.1   Single Layer Test Sample Thermal Profiles 45

4.1.2   Multi Layer Test Sample Thermal Profiles 56

4.2   Hardness Test Results 64

4.2.1   Hardness Measurement Validation Results 65

4.2.2   Test Sample Hardness Test Result Summary 68

4.3   Microscopy Results 74

4.3.1   Single Layer Microscopy 74

4.3.2   Multi-Layer Microscopy Results 78

Chapter 5.   Discussion 83

5.1   Thermal Modeling Versus As-Built 83

5.2   Hardness Testing 85

5.3   Microscopy 85

5.4   Traditional GTAW Processing Versus GTAW DED MAM 86

Chapter 6.   Summary 88

Chapter 7.   Future Work 92

7.1   Considerations for Modeling and Predictions of Present Phases 92

7.2   Applicability to Other AM Processes and Heat Treatment Resolution 93

References 95

Appendix A.   DATM Post Processing MATLAB Code.. 104

Appendix B.   DATM CCT Curve Generation of Model Overlay MATLAB Code....... 108

Chapter 1.   Introduction

Metal additive manufacturing (MAM) has matured since additive manufacturing's (AM) initial development as a rapid prototyping tool for plastics, with only 8% of production parts being manufactured with this process in 2004 (Bearman, Bourell, Seepersad, & Kovar, 2020) . Although rudimentary MAM dates to the mid 1920's, the primary transition to modern AM occurred in 1984 with the release of more readily available computational resources (Bourell, 2016) (Williams, et al., 2016) (United States Patent No. 1,533,300, 1925). As there are numerous MAM processes, selection of a suitable one for particular applications takes into account all the various aspects summarized on the diagram shown in Figure 1.1 (Gradl, et al., 2021). The motivation of this study is to develop the technology for realizing spatially resolved material properties when using MAM processes. As microstructural evolution and hence properties are dependent on the heat treatment applied, the ability to perform these *in-situ* rather than post- build would help realize one of the potential benefits of the MAM process.

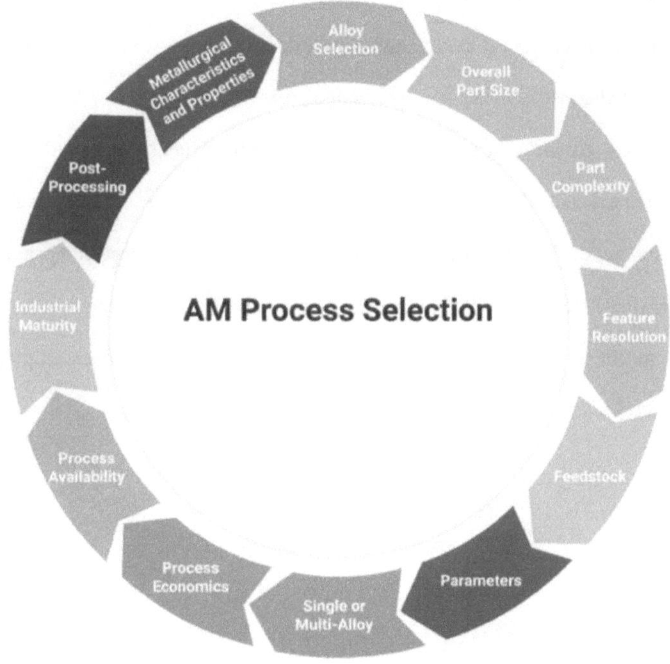

Figure 1.1: Additive Manufacturing Process Selection Overview (Gradl, et al., 2021).

The first objective of this study was to automate a commercially available Gas Tungsten Arc Welding (GTAW) process to realize *in-situ* heat treatments during an MAM process. GTAW is an arc-based heat source, similar to Gas Metal Arc Welding (GMAW), that lends itself to relatively inexpensive equipment relative to other direct energy deposition (DED) processes. In addition, GTAW provides the ability to decouple the heat from the feed source. The second objective was to utilize a numerical model to predict the boundary conditions and GTAW process parameters required to locally alter the microstructure during a build. The final objective relied on verifying the results from coupling the predictive modeling with the automated GTAW

DED process to obtain spatially discrete microstructures.

To validate the numerical model predictions, thermocouples were adhered to the base plate for real-time thermal monitoring of the temperature during the GTAW DED process. These thermal data are compared to the predicted thermal profile to ensure the intended heat parameters and timing were achieved. The results of the *in-situ* heat treatment are verified using hardness measurements along the length of the deposited weld. Hardness results are compared for heat treated and non-heat-treated samples in order to determine effectivity of the process. Finally, metallurgical specimens were prepared from the depositions, and optical microscopy performed to provide a visual verification of microstructural changes in the samples.

Chapter 2. Background

2.1 Additive Manufacturing Processes

Additive Manufacturing (AM), as defined in ASTM F2792 (ASTM, Standard Terminology for Additive Manufacturing Technologies, 2012) is the process of depositing material to additively make objects from 3D model data and is typically done in a layer-by-layer fashion. In general, metal AM processes are categorized by three major components: Material Introduction, Heat Source, and Feed Stock. In a metal AM build, this layer-by-layer addition of material results in a complex thermal history which may vary throughout the part based on part geometry (Zheng, Zhou, & Smugeresky, 2009). The changing heat profile results in non-uniform, or heterogenous, microstructures (Navarro-Lopez, Hidalgo, Sietsma, & Santofimia, 2017). Post processing, *ex-situ* heat treatments are designed to homogenize the build material, resulting in uniform mechanical properties. Thus, the post build heat treatments of AM parts have focused on uniformity and not obtaining spatially resolved mechanical properties (Schneider, 2020). Post process, *ex-situ* heat treatments increase the overall processing time and hence, cost. It also doesn't realize one of the potential benefits of AM processing, that of obtaining site-specific material properties (Tammas-Williams & Todd, 2016), which is the focus of the present study.

Figure 2.1 illustrates three main categories of MAM that can be broadly broken into: Powder-bed, Powder-fed, and Wire-fed. The categories are further detailed in Figure 2.2, where the AM processes are broken out based on feedstock and fusion technique (Gradl, et al., 2021).

The selection of the feed stock typically trades off feature resolution, build time, and part size.
With powder-bed based processes, the size of the part is confined to the size of the powder bed
although it allows for the finest resolution of features. To utilize AM for larger components, the
powder fed and wire fed processes can take place outside of a box and are referred to as directed
energy deposition (DED). Over time the nomenclature for the processes have evolved for
standardization (Frazier, 2014). For example, many of the early wire fed processes that were
originally referred to as shaped metal deposition (SMD) or wire arc (WAAM), are now
standardized within DED processing. Figure 2.2 illustrates how the feed stock and heating
source are used as a prefix to DED to identify the process.

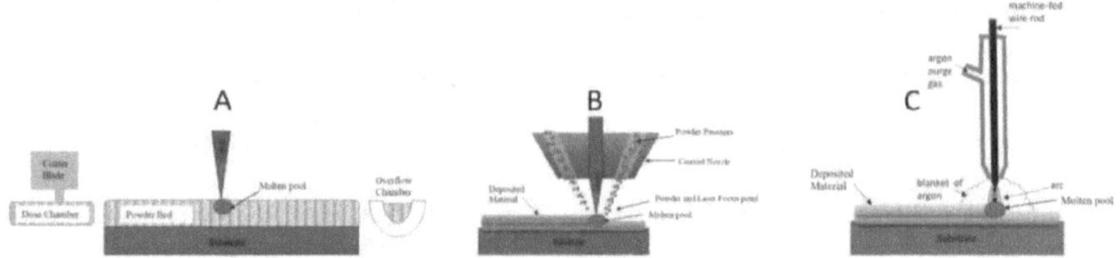

Figure 2.1: MAM Process Overview A) L-PBF, B) LP-DED and C) AW-DED.

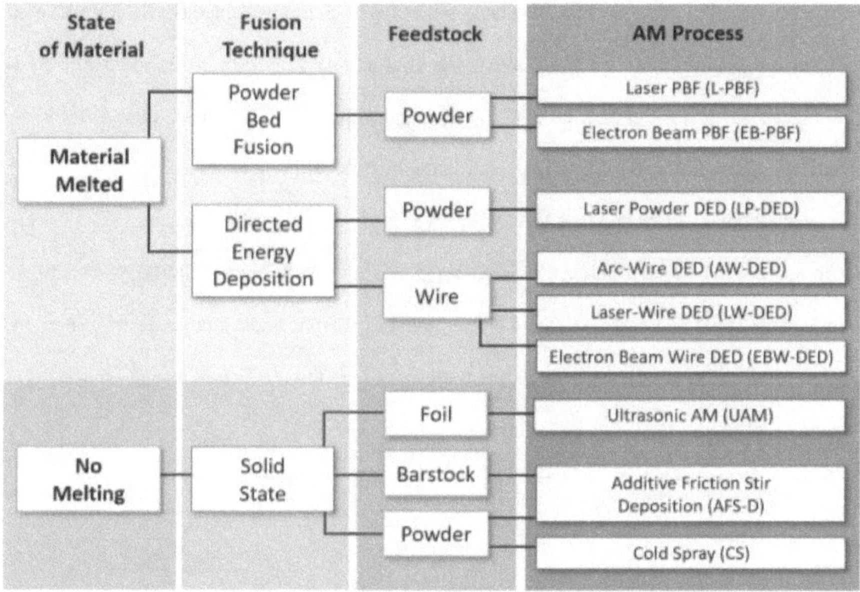

Figure 2.2: AM Process Diagram Based on Feedstock and Fusion Technique (Gradl, et al., 2021).

2.1.1 DED Processing Using Wire Feed

In DED processing, the heat and feed source are coupled with their positions dictated by the geometry of the component being manufactured. As noted in Figure 2.2, DED processes can use either a laser, arc or electron beam (EB) heating source. For arc-based DED, an arc is generated electrically and provides the heat to melt and join metals. Low cost methods for arc based DED have primarily focused on adapting the GMAW welding process (Pattanayal & Sahoo, 2021) (Williams, et al., 2016).

Wire fed processes are characterized as having the highest deposition rate among AM processes at the expense of feature resolution and surface finish (Gradl, et al., 2021).

Additionally wire fed processes are associated with feedstock that is inexpensive (Costello, et al., 2023). Thus, these processes are used more in the production of large structures to minimize machining waste (Schneider & Gradl, 2022). Mechanical properties of wire fed processes typically fall between those found in cast and wrought material (Wu, et al., 2018).

A qualitative comparison of MAM processes is given in Figure 2.3 adapted from Garcia-Colomo and ASTM-F3187 (Garcia-Colomo, Wood, Martina, & Williams, 2020) (ASTM, 2016). In comparison, the AW-DED (aka WAAM) process provides the highest build/deposition rate and unconstrained build volume while at the trade-off of the lowest feature resolution and dimensional accuracy.

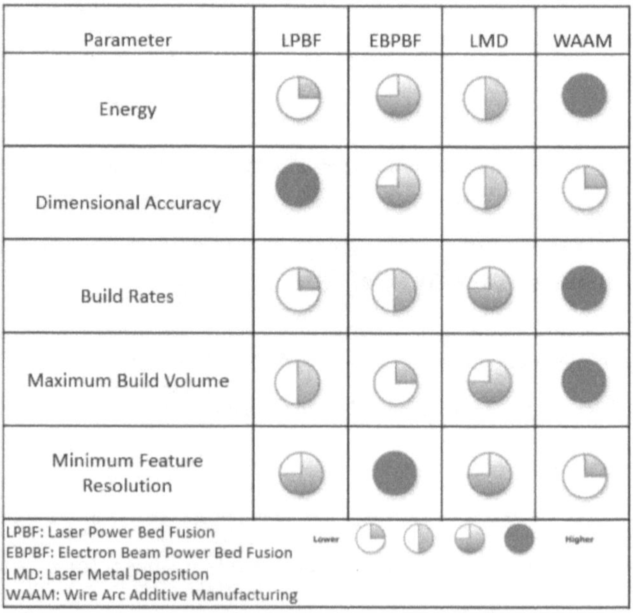

Parameter	LPBF	EBPBF	LMD	WAAM
Energy				
Dimensional Accuracy				
Build Rates				
Maximum Build Volume				
Minimum Feature Resolution				

LPBF: Laser Power Bed Fusion
EBPBF: Electron Beam Power Bed Fusion
LMD: Laser Metal Deposition
WAAM: Wire Arc Additive Manufacturing

Lower ◔ ◐ ◕ ● Higher

Figure 2.3: Qualitative Comparison of Product Results for Various MAM Processes – Adapted from Garcia-Colomo and ASTM-F3187 (Garcia-Colomo, Wood, Martina, & Williams, 2020) (ASTM, 2016).

In contrast to fusion based welding intended to join two workpieces, the arc-based wire fed process can be adapted to AM processing by laying multiple weld beads on top of each other as the part is built. Figure 2.4 illustrates the GTAW process and can be contrasted with the GMAW process illustrated in Figure 2.1c. The primary difference between the two arc-based processes is that the wire feed is independent of the heat source in the GTAW process. A non-consumable tungsten electrode is connected to the weld control unit and serves as the negative lead, where the base material is connected with a clamp lead that serves as the positive signal of the circuit. Similar to GMAW, an argon shielding gas protects the molten material from contamination and oxidation during the welding process. The circuit is energized for a GTAW welding process by forming an electrical arc between the non-consumable tungsten electrode and base material. This arc can be maintained indefinitely, either statically or while the tungsten electrode is in motion, as long as the critical distance between the electrode and working material is not exceeded. The critical distance is dependent on the current (heat input) settings of the weld unit. While the maximum current is set on the weld unit, real time current changes are typically made using a foot-controlled petal. In industrial weld settings, the tungsten electrode is manually manipulated to perform the welding while the base material stays fixed. For automated welding processes, either the base material, electrode, or both are manipulated to join the workpieces of interest. Once a consistent arc has been produced, filler metal in the form of a wire is added to produce the weld bead. The filler material is selected to be compatible with the workpieces and the desired mechanical properties.

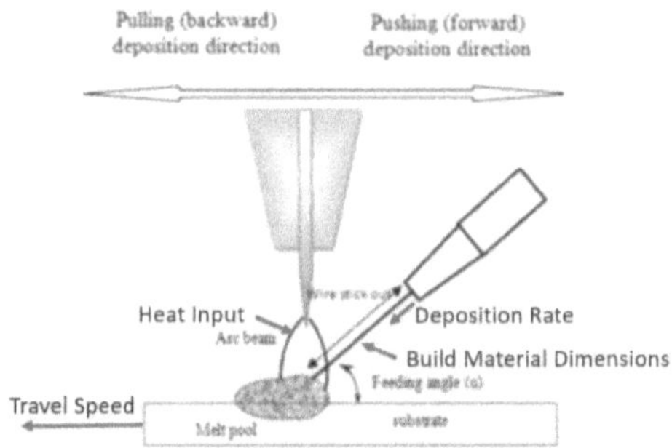

Figure 2.4: GTAW Welding AM Configuration with Heat Source Decoupled from Feedstock.

An advantage of GTAW is the elimination of spatter during the welding process that is common in GMAW welding, as shown in Figure 2.5 (Rodriguez, et al., 2018) . Formation of spatter is undesirable, as it affects the surface finish and can be entrapped in subsequent layers, reducing mechanical properties. During GMAW welding, the wire filler material serves as the consumable electrode that is continuously being eroded. This results in molten droplets which fall into the weld pool and can be ejected resulting in splatter. Alternatively, in a GTAW based process the filler material is added directly to the molten pool which prevents the spatter effect. Additionally, the GMAW process initiates the arc while simultaneously beginning to feed new wire into the weld pool. The GTAW process, on the other hand, generates the arc independently from the feed source.

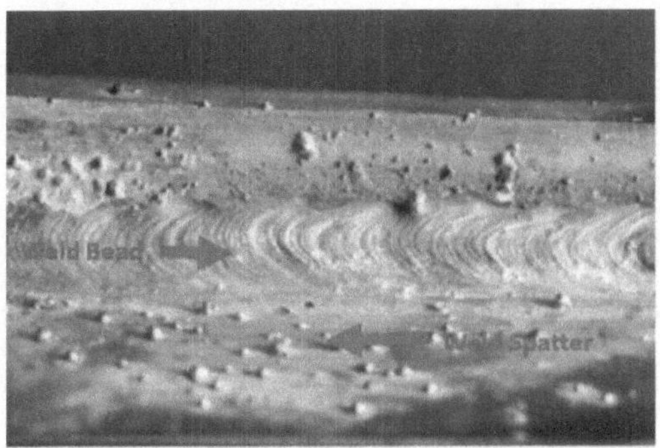

Figure 2.5: Weld Spatter Example from GMAW Welding.

2.1.2   GTAW AM DED Processing

The GTAW-AM process has been investigated for its potential benefits on mechanical and geometric improvements as compared to other arc wire processes (Gokhale, 2019). Previous investigations into the use of a GTAW based AM process have focused primarily on process parameters to optimize the mechanical (*i.e.,* layer adhesion, deposition density) and geometric (*i.e.,* feature resolution and tolerance/accuracy) characteristics of the deposition. Making use of the uncoupled heat source, researchers have studied the impact of feedstock to torch orientation on the dimensional and mechanical properties of thin-walled structures (Gokhale, Kala, Sharma, & Palla, 2020) (Geng, Li, Xiong, Lin, & Zhang, 2017). It was ultimately determined that forward wire feeding with a low wire incidence angle produced the best combination of geometric and mechanical properties. In addition to variability in the wire feed orientation, studies have also been conducted on the optimization of the arc length and wire feed rate in an effort to maximize feature resolution (Geng, Li, Xiong, Lin, & Zhang, 2017) (Liu, Feng, Chen, & Chen, 2023). Instead of controlling the feed orientation, the feedstock can be introduced coaxially (Rodriguez, et al., 2018). This allows the feedstock to torch orientation to remain constant independent of travel direction and introduces the feedstock into the center (and hottest) portion of the arc which can allow for a reduction in heat input to the part.

Arc wire processes, including GTAW based AM, have been used for fabricating pyramidal lattice structures in stainless steels through varying of the heat input and layer height (Zhang, et al., 2020). The use of GTAW based AM with a pulsed power source has been also been used to develop unsupported angled lattice structures by reducing heat input with a pulsed arc (Xu, et al., 2020). The feedstock for the GTAW based MAM process has been pre-heated prior to entering the established arc, requiring a lower heat input from the GTAW source and reducing the thermal input into the deposition (Li, et al., 2019). This process was used to alter the grain structure in the initial deposition due to the ability to control heat input.

Combining the optimized wire feed orientation and arc length, numerical models have been developed to predict the deposition width using a single wire feed orientation with multi-directional processing such that the wire feed speed and arc length can be altered *in-situ* to maximize geometric tolerancing (Wang, Wang, & Li, 2020). Process parameters have also been reviewed for their respective impact on deposition quality during the GTAW process. The process current, whether AC or DC, and the deposition speed were used as inputs into numerical models to predict impacts on the resulting deposition geometry (Cai, Dong, Lin, Li, & Fan, 2022) (Chen, Du, Zhang, & Zhao, 2023). *In-situ* arc length control and wire feed orientation have been reviewed experimentally for geometric and mechanical repeatability of the deposition (Wang, Wang, Wang, & Li, 2019).

Although significant advancements have been made in arc wire based AM, the full potential of the GTAW process has not been explored. One such potential is the ability to use the decoupled GTAW based heat source to be adapted for *in-situ* heat treatment as part of the process parameters. Thus by addressing this limitation in the current literature, this study would further expand the use of arc based, wire fed AM to include GTAW as a viable method for spatially controlled microstructures.

2.2 Numerical Modelling of Heat Profile

Implementation of the arc based wire processes into AM, that were shown in Section 2.1, can lead to time consuming trial-and-error to develop and validate process parameters for each process and build geometry (Stockman, 2019). The ability to numerically predict the heat profile relies on the flux from the heat source to be used as input. As the flux is difficult to directly measure in arc based processes, it is usually calculated starting with the heat input (power) per unit length as given in Equation 2.1:

$$q = \frac{V * I * \eta}{s}, \tag{2.1}$$

where q is the power, V is the voltage, I is the current, η is the weld (heat source) efficiency, and s is the travel speed. In order to determine the heat flux, which is the heat input per area, Equation 2-1 is divided by the width of the arc beam show in Figure 2.4. The final calculation of heat flux is given by (Wu, et al., 2018) (Kumar, Gautam, & Kumar, 2014) in Equation 2.2:

$$Q = \frac{V * I * \eta * 60}{s * 1000 * w}. \tag{2.2}$$

Where Q is the heat flux, $V*I$ results in Watts, the factor of 60 converts minutes to seconds, the factor of 1000 converts joules (J) to kilojoules (kJ) and dividing by the arc beam width (w) results in kJ/Area.

As the calculations in Equation 2.1 and 2.2 depend on the welding efficiency, this information can be obtained from published literature (Kou, 2021), similar to that shown in Figure 2.6 , and is typically around 60% for GTAW. The efficiencies (η) of the selected heat source are used in conjunction with the voltage and current settings of the welder and Equation 2.2 to determine heat flux (Q) during deposition.

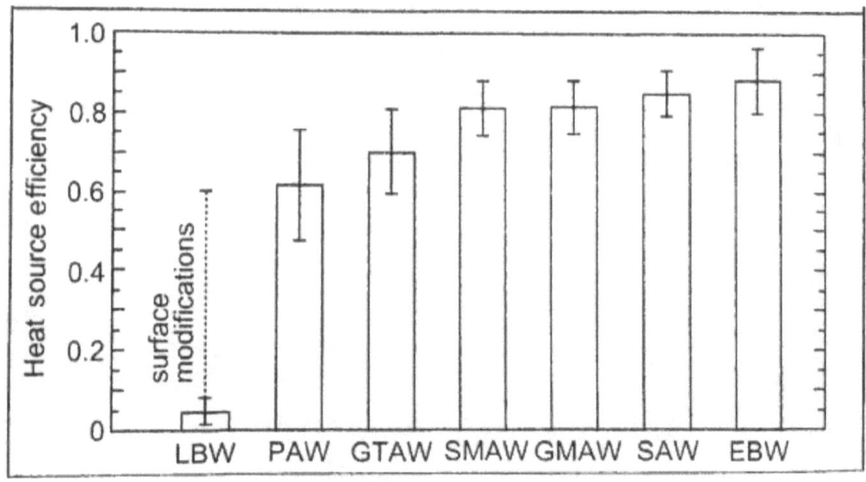

Figure 2.6: Heat Source Efficiencies in Various Welding Processes (Kou, 2021).

Various researchers have used computational thermal models to develop MAM process maps (Beuth, et al., 2013) (Gockel & Beuth, 2013). There are four primary modeling fields, shown in Figure 2.7. Each field is concerned with different physics and length scales that occur during the MAM process (Francois, et al., 2017). Additionally, the timescale for the models are dependent on both axis. As the length scale increases, the time for process changes to occur also

13

increases therefore the required timescale increases. Similarly, rapid thermal changes occur when observing microstructural evolution (microstructural modelling) vs properties modelling which can occur in a quasistatic manner. Microstructural modelling is primarily focused on materials and cooling rates along with the associated microstructure, (Karma & Tourret, 2016) (Hoyt, Asta, & Karma, 2003), while properties modelling focuses on material behavior at a small length scale stemming from the microstructure (Lim, et al., 2016). Process modelling is used for predicting thermal and residual stress resolution in the entire build and base plate, (Michaleris, 2014) (Martukanitz, et al., 2014), while performance modelling focuses on response of the part relative to its intended use (Francois, et al., 2017). One of the more significant obstacles in AM process modelling is the required computational expense.

A less computationally extensive numerical model has also been developed for predicting the global temperature gradient in a build. Inputs to the model included the deposition parameters, or heat flux, as a function of the part geometry (Stockman, 2019). This approach was labeled a Finite Difference Additive Thermal Model (DATM) and used a finite difference method with mass added to predict the global temperature distribution during an AW-DED build (Stockman, 2019). This computationally efficient, numerical model provides a first order prediction for the global temperature profiles.

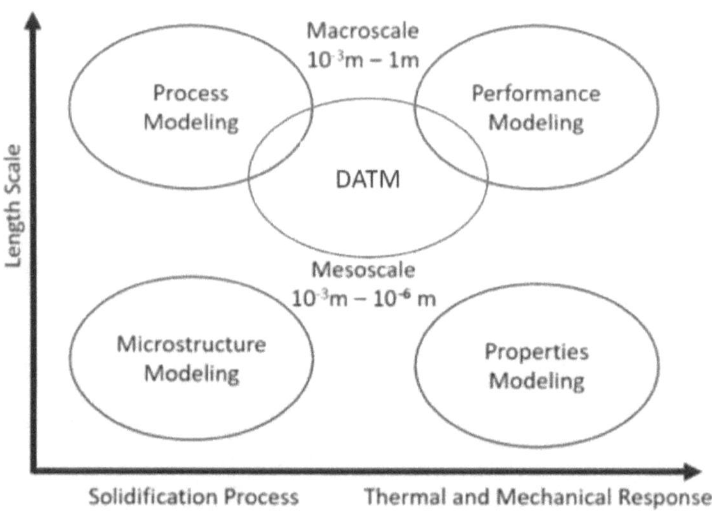

Figure 2.7: AM Modeling Categories Relative to Length Scale and AM Process of Interest.

The DATM model makes use of simplifying assumptions of the complex AM processes to achieve computational efficiency. Heat flux is constrained to a Cartesian mesh cube located around the center of the added mass (molten pool). Mesh geometry is defined by the user, where the entire part is meshed during computation, but nodes are only computationally active after mass has been added at the node location. Conduction serves as the main source of heat transfer, with convection also included. As the effect of radiation was found to be only a minor heat transfer contributor, it is only considered for material at a temperature above 980°F (Stockman, 2019). The convection term becomes more prominent as the exposed surface area of the build increases. The DATM model allows the user to select points of interest for plotting their time history temperature data. The data can be exported to allow for more precise data probing, beneficial for comparing with AM processes parameters. The predictive capabilities of DATM have been verified in AW-DED studies on a 4340 steel alloy (Stone, 2020).

2.3 Heat Treatment and Microstructure

The mechanical properties of a metal are controlled by their microstructure, which is influenced by the processing route. In ferrous alloys, various non-equilibrium microstructures can be produced based on the quenching used in the heat treatments. An earlier form of MAM was used in cladding operations onto worn surfaces using either a wire or powder feedstock. For laser cladding processes, selective control of the laser beam shape (Shang, et al., 2014) and temperature profiles between cladding layers (Sun, et al., 2018) was shown to increase uniformity in the final microstructure. As noted in Chapter 1, a MAM part experiences multiple thermal cycles where each thermal cycle may have varying cooling rates. Thus, it is important to understand the initial temperatures and resulting cooling rates to actively control the formation of the desired microstructure and resulting properties.

2.3.3 Equilibrium Conditions

Figure 2.8 shows the steel portion of a Fe-C phase diagram and its equilibrium phases that form when the metal is heated above the eutectoid temperature and allowed to slowly cool. From the phase diagram in the hypoeutectoid region, it can be seen that during cooling at starting temperatures above the A3 temperature, the single austenitic γ-Fe phase decomposes into two phases of ferrite (α) and cementite (Fe_3C) (Honeycombe, 1995). All heat treatment of ferrous alloys thus relies on solutionizing the alloy above the A_3 temperature into the γ-Fe single phase region and then controlling the cooling rate during the decomposition.

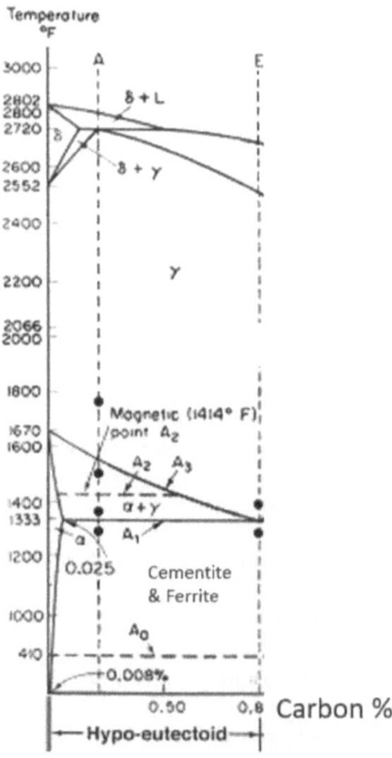

Figure 2.8: Fe-C Equilibrium Phase Diagram Hypo-eutectoid Region (Honeycombe, 1995).

Austenite (γ) has a face-centered cubic (FCC) lattice structure with higher carbon solubility, whereas α has a body-centered cubic (BCC) lattice structure with lower carbon solubility. For hypoeutectoids, as the temperature is decreased from A_3 to A_1 the metal undergoes a phase transformation from the single-phase region into the two-phase region of α + γ. The α that forms initially is referred to as the proeutectoid α. As the temperature further decreases below A_1, the eutectoid temperature, the remaining γ-Fe decomposes into cementite + α forming a lamellar structure of alternating layers. The lamellar structure is often referred to as

17

pearlite and its hardness is controlled by the lamellar layer thickness. Slow cooling results in the largest lamellar spacing and lowest strength, whereas increasing the cooling rate forms smaller lamellar spacing and higher strength. The strengthening mechanism relies on interfacial strengthening between the α and Fe_3C phases (Kinney, et al., 2017).

2.3.4   Non-Equilibrium Conditions

The equilibrium conditions of Section 2.3.3 only apply to slow, equilibrium cooling rate conditions. Thus, the effects of cooling rate and the impacts on the resulting microstructure cannot be determined from the equilibrium phase diagram. To control the non-equilibrium phases, an isothermal time-temperature-transformation (TTT) is used to guide the selection of times and temperatures for formation of the various phases. Figure 2.9 shows a representative TTT diagram for the hypoeutectoid steel, SAE 4340. Increasing the cooling rate can result in either pearlite or bainite formation. If the γ can be retained to lower temperatures, bypassing the nose of the transformation curves, martensite can form with its body centered tetragonal (BCT) lattice structure. By rapidly cooling the austenite, the carbon atoms become trapped in the FCC structure thereby forcing a diffusion-less transition to a BCT lattice structure (Honeycombe, 1995). One method to evaluate the final phases formed is by the corresponding hardness. Of the ferrous alloy phases, martensite has the lowest ductility and highest strength corresponding to hardness, as shown on the right axis of Figure 2.9.

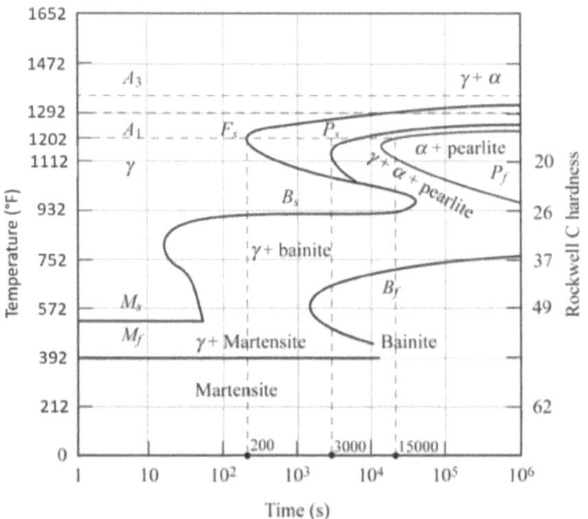

Figure 2.9: SAE 4340 Time-Temperature-Transformation (TTT) Diagram – Grain Size: 7-8 (ASM International, 1995).

One of the limitations of TTT diagrams is that they are based on isothermal processing. In practice, heat treatment operations are not carried out isothermally but are carried out where cooling rates vary as a function of material thickness due to differences in temperature between the surface and interior. Because of this, continuous cooling transformation (CCT) diagrams, shown in Figure 2.10 for SAE 4340, address the varying cooling rates leading to the development of mixed phases (Kou, 2021) (Navarro-Lopez, Hidalgo, Sietsma, & Santofimia, 2017).

Because AM processes introduce rapid and cyclic heating and cooling and the associated temperature gradients, a CCT curve provides valuable insight for predicting final microstructure. Figure 2.11 shows a comparison of the relative thermal peak and time history of a MAM deposition compared to traditional heat treatment along with a condensed Fe-C phase diagram

19

for comparing to the temperature required for phase transformations.

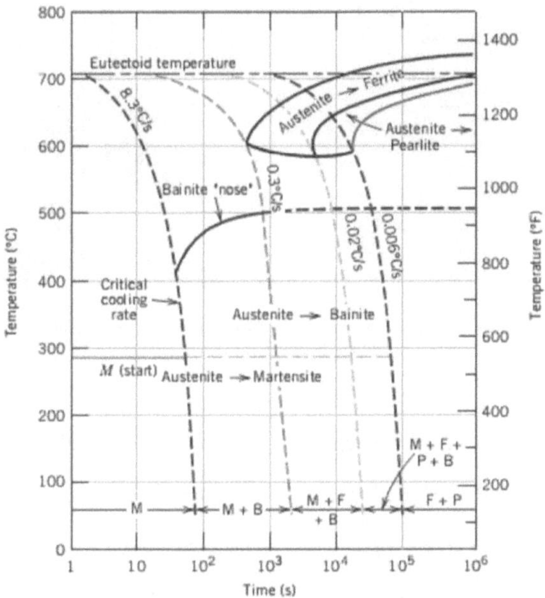

Figure 2.10: SAE 4340 Continuous Cooling Transformation (CCT) Diagram – Grain Size: 7 (ASM International, 1977).

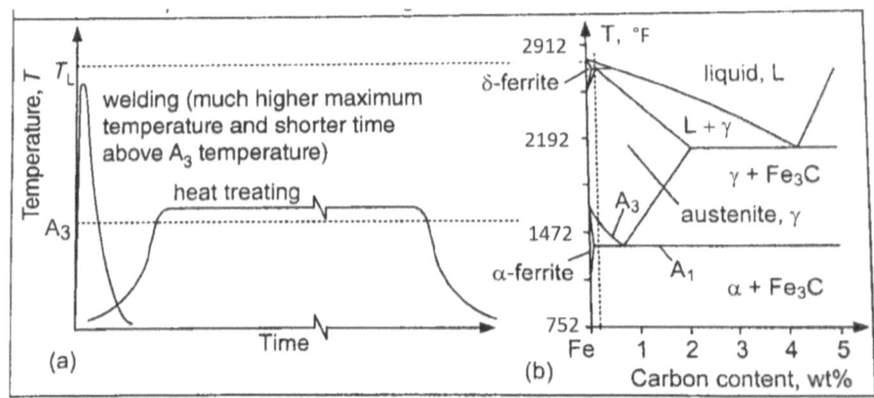

Figure 2.11: Traditional Heat Treatment Compared to AM Deposition Thermal Conditions (a) and Comparative Fe-C Phase Diagram (b) (Kou, 2021).

As shown in Figure 2.11, the deposition process results in heating of material adjacent to the deposition between the A_3 temperature and the liquidus temperature. Also noted is the shorter cooling time to transition from the peak temperature down to ambient temperature that corresponds to a faster cooling rate. In contrast, a traditional heat treatment would be applied to the material for longer times above the A_3 temperature to ensure complete transformation to the γ phase prior to decomposition. The time held at temperature would be dependent on the thickness of the part to ensure a uniform temperature was reached and phase transformation to γ was complete. As mentioned previously, the phase transformations are diffusion based (with martensite being an exception) and the timing is dependent on the material mass, diffusion rate kinetics, and the cooling rate. With rapid heating during deposition, the A_1 and A_3 temperatures can increase resulting in an adjusted Ac_1 and Ac_3 temperatures, respectively (Mostafaei & Kazeminezhad, 2016) (Deng, Li, Di, & Misra, 2019). The impacts of heating rate on austenite formation have been studied previously revealing that austenite formation and grain size are heat

21

rate dependent. As such, prediction of the microstructures present during cooling assume complete austenite formation (Cerda, et al., 2016). Ultimately, the hardenability of the material is also austenite grain size dependent where increasing austenite grain size results in increased hardenability due to the grain boundary are per unit volume decreasing (Bhadeshia & Honeycombe, Steels: Microstructure and Properties, 2017). The need for new reference temperatures for phase transformation (Ac_1 and Ac_3) is due to the short duration above the A_3 temperature that reduces the availability for complete austenitic phase transformation (Kou, 2021). The reduction in diffusion due to decreased time is partially offset by increased diffusion occurring at the much higher temperatures. For a weld based process the average martensite to austenite transformation rate increases with increasing heat rate (up to 212 °F/s), while the maximum transformation corresponds to the lowest heating rate (Zappa, Hoyos, Tufara, & Svoboda, 2022). This is due to the increased heating rate resulting in a higher maximum temperature, but a faster heating rate reducing time for austenite formation. Adjusted CCT diagrams for heating of material due to welding can be obtained with a static heat source and a dilatometer used to detect the resulting volume changes corresponding to phase transformations. However, these curves are often not available and instead standard CCT curves are often used (Kou, 2021). In additional to the impacts of austenite grain size due to intentional or un-intentional heat cycling has shown that a reduction in prior austenite grain size promotes retained austenite during cooling (Hidalgo & Santofimia, 2016). Other researchers, (Obasi, et al., 2019) (Lee, 2013) (Jeon, et al., 2022), have developed analytical models for predicting austenitic phase transformations during rapid heating and subsequent transformations during cooling due to weld-like thermal conditions of the material adjacent to the weld and not exceeding the liquidus temperature. However, this study assumes that reheating of material (below the liquidus temperature) due to additional deposition layers produces sufficient austenite transformation to use the nominal CCT curve for predicting measurable phase decompositions. The impact of this assumption is that any effects from martensite tempering or retained austenite in the final microstructure are not considered. Further refinement of austenite formation modeling is discussed in Chapter 7.

2.3.5   Effects of Thermal Cycling

During the build process, each layer is deposited onto the subsequent layer, exposing the previous layer to an increase in temperature followed by a cooling rate that is dependent on the part geometry and thermal boundary conditions. Example data from a MAM process are provided in Figure 2.12, which captures the thermal history at a fixed point in the baseplate that is consistent for each trace. For a single trace, each peak corresponds to the resulting temperature change in the base plate as an additional layer is added in the build. As the build progresses the temperature delta at the point of interest when the heat source passes decreases, while the average temperature at the point increases. Each trace represents the thermal response of the same point in the build plate with in an increased initial temperature. Thus, if the substrate temperate is elevated (potentially due to previous processing or intentional pre-heating) the critical number of layers where the point of interest exceeds the A_1 eutectoid temperature will be affected. Three main points are obtained from Figure 2.12. First, the thermal cycling has been shown to provide an unintentional *in-situ* heat treatment of previous layers (Stone, 2020). Heat input experienced during subsequent layer depositions can increase the temperature at the point of interest above the eutectoid temperature resulting in the unintentional heat treatment. In Figure 2.12, at least three layers above the initial deposition increased the temperature beyond the eutectoid temperature. Second, the figure shows that if intentional *in-situ* heat treatment is

23

desired the user must be selective on when in the build the heat treatment is applied to ensure that subsequent layer deposition does not alter the applied *in-situ* heat treatment. Applying an intentional *in-situ* heat treatment prior to the critical number of layers being reached could result in unintentionally altering the desired heat treatment due to exceeding the eutectoid temperature. Finally, Figure 2.12 shows that the boundary conditions, substrate temperature in this case, affect the temperatures that will be experienced in the deposition. Thus, boundary conditions must be taken into consideration when determining how and when the *in-situ* heat treatment will be applied.

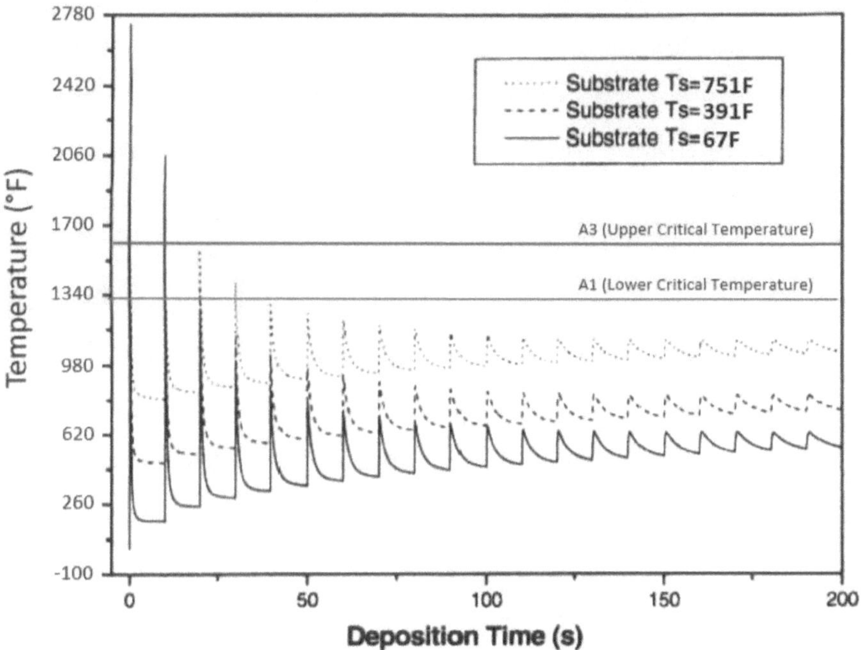

Figure 2.12: Multi Pass Example Temperature Time History of a Fixed Point in the Baseplate with Varying Initial Substrate/Baseplate Temperatures (Zheng, Zhou, & Smugeresky, 2009).

Chapter 3.   Experimental Methodology

3.1 Numerical Modeling

To associate the GTAW processing parameters with the global heat profile, the DATM numerical model was used. Results from the DATM predictions are used to determine weld parameters for a given test sample (*i.e.,* current settings, travel speed, number of passes, etc.). That is, the model provides the heat input and timing required for nominal deposition or *in-situ* heat treatment. Thus, model results needed to be validated and verified against test data. For a given model the only parameter that is derived instead of directly implemented is the heat input from the simulated heat source. To validate the model predictions, a weld pass is executed that matches model geometry with a welder current setting selected to most accurately replicate the heat input used in the model. The model validation weld pass does not introduce any feed material to reduce variables present for model validation. The weld parameters used during the model validation replicate those that are used during the actual deposition process. Figure 3.1 shows the temperature profile of the thermocouple located on the build plate compared to the profile predicted by the DATM model. Thermocouple placements for all samples are similar, where the thermocouple is placed on the edge face of the build plate centered on the build plate length (3") and centered about the height of the build plate (1"). A typical thermocouple placement with dimensions is provided in Figure 3.2, and any deviations for each sample are provided in Section 4.1.

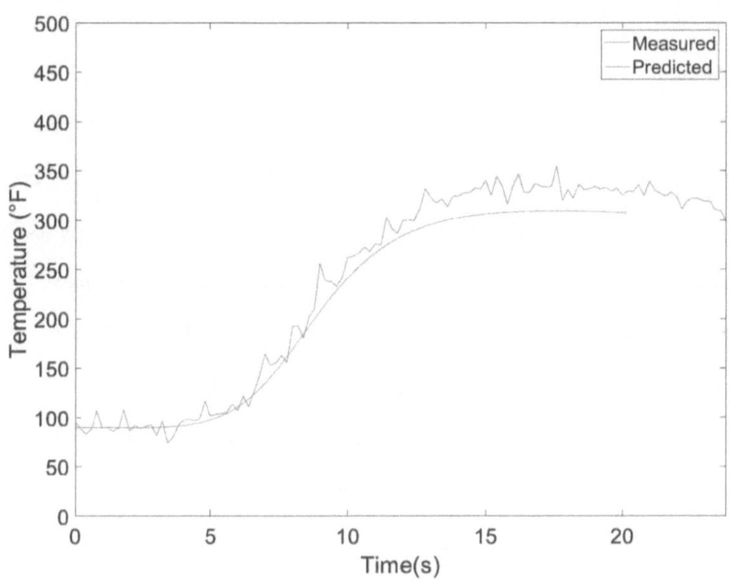

Figure 3.1: Model Validation Prediction Versus Measured Example.

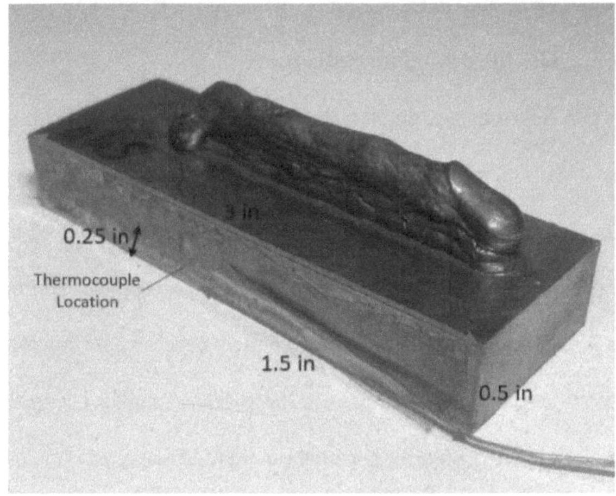

Figure 3.2: Typical Thermocouple Placement Relative to Build Plate Dimensions.

The validated DATM model is used to predict the cooling rates for various

microstructures. This guides the location of using the decoupled heat source to locally provide

extra heating and the required durations for the applied heat. The DATM program only

considers the effect of varying heat flux and thus doesn't capture details of the actual GTAW

parameters. Two models in DATM are generated, one to validate the heat flux for the initial

deposition layer and the other to validate the heat flux for the *in-situ* heat treatment.

Initial deposition produces a molten pool of metal followed by solidification with rates on

the order of 600 °F/s, which are faster than the critical cooling rate in Figure 2.10 of 47 °F/s, due

to the boundary conditions being at a temperature of 80 °F (ambient) and the low thermal mass

of the molten pool (4 x 10^{-4} lbs.) relative to the adjacent build material (0.43 lbs.).

For the *in-situ* heat treatment, the GTAW parameters are modified to introduce a lower

heat flux to avoid remelting of the material. The DATM model is then used to determine the heat

input settings, dwell times, and number of passes required for *in-situ* heat treatment. This is done

by comparing the model output to the CCT curves of Figure 2.10 and adjusting the heat flux until

the resulting curve provides the desired microstructure.

For the 10-layer heat-treated samples, there is not a separate DATM model generated.

Instead, predictions use a superposition of the 10-layer non-heat-treated model and the single

layer heat-treated model. That is, the 10-layer non-heat-treated model is used to identify which

layer during the deposition process the heat treatment should be applied, and the single layer

heat-treated model is used to determine the heat treatment processes setting. This approach can

only be implemented due to the heat treatment in the 10-layer samples occurring at a point in the

build where the build plate still primarily controls the thermal response. This can be seen because

the maximum possible thermal mass of the deposition at time of heat treatment is 0.016 lbs. and the build plate thermal mass is 0.43 lbs.

3.2 Material

A 4340 hypoeutectoid ferrous alloy was selected for this study due to the known heating

and quenching influence on resulting microstructures. SAE 4340 is used in commercial and

military aircraft, automotive components, and firearms where high strength and fatigue

resistance properties are desired. Table 3.1 summarizes the chemical composition of the 4340

and the equivalent ER4340 wire feed stock. The diameter of the ER4340 used was 0.0625 in and

did not contain flux additive. Both SAE 4340 and ER4340 are known for their toughness and

ability to attain high strengths after heat treatment. Additionally, 4340 is known for its good

fatigue resistance before and after heat treatment

Table 3.1: Chemical Composition of SAE 4340 (ASM International, 1993) and ER4340 (WA. Alloy Corporation).

Element	SAE 4340 Wrought Material Composition (wt %)	ER4340 Material Composition (wt %)
Iron, Fe	95.195-96.33	Balance
Nickel, Ni	1.65-2.00	1.80
Chromium, Cr	0.700-0.900	0.78
Manganese, Mn	0.600-0.800	0.85
Carbon, C	0.380-0.430	0.35
Molybdenum, Mo	0.200-0.30	0.25
Silicon, Si	0.200-0.350	0.50
Sulfur, S	0.0400	0.014
Phosphorous, P	0.0350	0.011

The chemical composition of ER4340 matches closely in terms of intentional alloying

elements, although ER4340 has additional Si content. Wrought SAE 4340 material properties are given in Table 3.2. The baseplate for all samples used in this study was low carbon (mild) steel.

Table 3.2: Mechanical Properties of Annealed 4340 (ASM International, 1993).

Property	Metric	Imperial
Tensile Strength	745 MPa	108 ksi
Yield Strength	470 MPa	68.2 ksi
Elastic Modulus	190-210 GPa	$27.6 - 30.5 \times 10^6$ psi
Elongation at Break	22%	22%
Hardness, Rockwell B	95	95
Hardness, Rockwell C	17	17

3.3 Weld Process Control Methodology

A Vulcan Pro TIG 250 GTAW welder was used as the power source for generating the builds. This welder was selected because it provides a 95 percent duty cycle at 110 amps, providing approximately 9.5 minutes of weld time, followed by a 30 second required cool-down window. The allowable weld duration is greater than the maximum two minutes required to complete a single weld pass. Additionally, the welder offers a current range up to 205 amps, exceeding the anticipated 100 +/- 20 amp need, and high-frequency start capability. The high-frequency start is vital to the test setup, as it allows the weld arc to be initiated without the tungsten electrode making contact with the base material. Thus, there is no need for active control of the vertical (Z) axis during weld deposition. The welder uses a direct current (DC) output with the tungsten electrode being negative (ground) of the circuit, *i.e.,* DCEN. Maximum current is set using the user interface with the applied current adjustable throughout the AM process. The AM process utilizes the maximum current set on the welder throughout the weld process to ensure repeatability of the weld samples. The build plate, with dimensions of 3" x 1" x 0.5" (Length x Width x Height), serves as a notable heat sink relative to the AM layers. Thus, as the build progresses, less heat is required to maintain AM layer geometry and adhesion to the previous layer. Similarly, excess heat can lead to distortion of the AM part geometry. As the build progresses, current is reduced to account for the changing boundary condition. Magnitudes of the current changes were determined via bench tests where layers were added at incrementally reduced currents until the weld geometry became non-uniform. The samples were sectioned to verify uniform adhesion between layers. Table 3.3 provides the weld current for each layer in the ten-layer builds. For adjustments that were made between layers, the current was adjusted prior to beginning the next weld layer. The baseplate for the builds is held in a metallic vise that is directly connected to the positive lead of the GTAW source to ensure proper contact is maintained.

Table 3.3: Per-Layer Weld Sample Current.

Layer Number	Weld Current (Amps)
1	110
2	110
3	110
4	100
5	100
6	100
7	100
8	95
9	95
10	95

3.4 GTAW Based MAM Platform

The GTAW source is mounted on a gantry for control of the deposition geometry. To adapt the GTAW welding for AM builds, two modifications were needed. The first was the high frequency start capability to initiate a weld without making contact with the substrate, and the second was customizing a system for automated feedstock delivery. The GTAW electrode holder is mounted on the X-axis of the gantry, leaving only the Y-axis and Z-axis as degrees of freedom in the AM depositions. Since the GTAW process decouples the feed and heat source, as discussed in Section 2.1, a separate feed device is required. To ensure repeatability between samples, the feed source needed to be able to accurately control the filler quantity and speed during deposition. Build material is fed through a custom designed feed source, shown in Figure 3.3, to allow for automated delivery. The feed angle was set at 8° with the base plate, based on prior research done by Gokhale (Gokhale, Kala, Sharma, & Palla, 2020) and Rodriguez (Rodriguez, et al., 2018). The feed source was designed with two rollers, each with an impression of approximately half the diameter of the feed material. This serves to guide the material and ensure proper contact while in operation. One roller serves as an idler, and is not actively driven, while the other roller is driven by a stepper motor. The drive commands for the feed source stepper motor come from the same control board as the Y axis so that the movement of the electrode and introduction of feed material can be linked. Each weld bead is initiated by creating an arc between the electrode and the previous layer of material. Once the arc is established a command is sent for the feed control to execute the previously programmed sequence of travel distance, travel speed, and feed quantity. These parameters for the nominal deposition and *in-situ* heat treatment are provided in **Table 3.4**. Argon served as the shielding gas during deposition with a flow rate of 10 ft^3/hr.

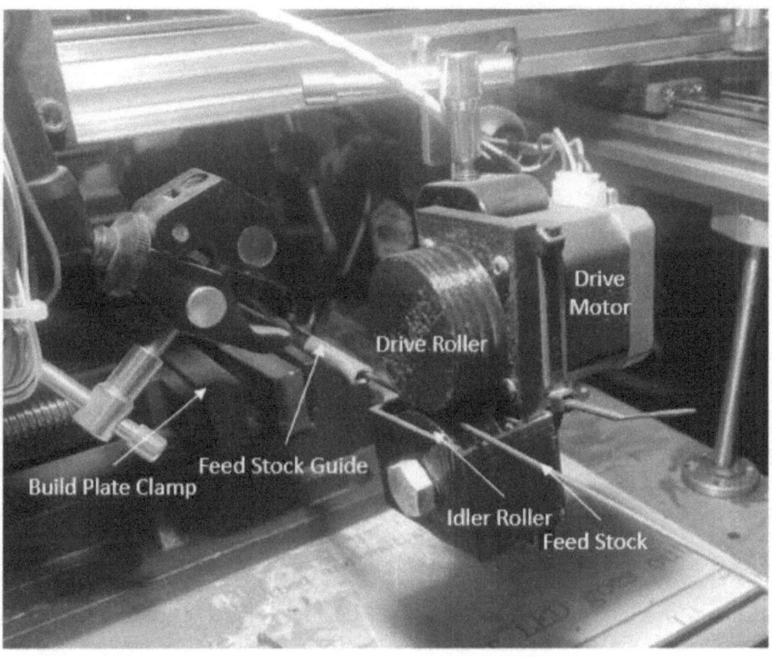

Figure 3.3: Feed Source and Associated Components.

Table 3.4: Deposition and Heat Treatment Gantry Inputs.

Deposition Type	Travel Distance	Travel Speed
Layer Deposition	2 in	7.8 in/min
Single Layer *In-situ* Heat Treatment	1 in	1.2 in/min
Multi-Layer *In-situ* Heat Treatment	2 in	1.2 in/min

3.5 Control System

Control commands are sent wirelessly over the local network. The control PC utilizes a 3rd party web interface for generating commands and G-code to prevent a radio frequency (RF) susceptible connection between the PC and gantry. A network diagram is provided in Figure 3.4. These commands are received by a Raspberry Pi, which is directly connected to the gantry and feed source. Wireless communication between the host PC and printer is critical to prevent detrimental electrical interference caused during the high frequency start process. Although considerably less susceptible, there still exists potential for disruption in connection caused by the high frequency start. Thus, a pre-check is conducted in which a high frequency start is initiated without the addition of filler material to ensure that loss of communication will not occur once the actual deposition is initiated. Initiation of the arc is manually triggered in conjunction with sending commands to the gantry and feed source. Once initiated, the process is automated.

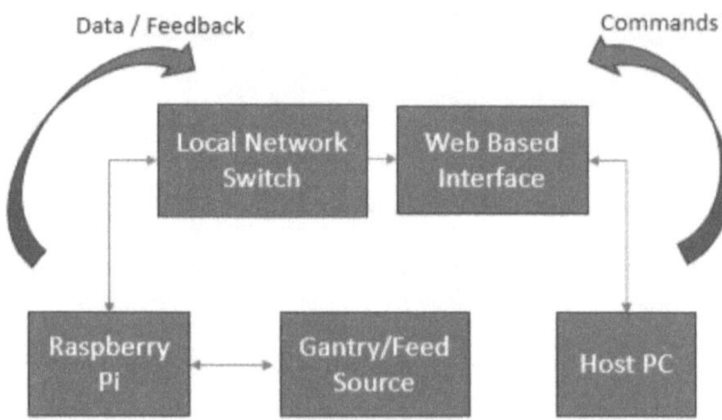

Figure 3.4: Control System Network Diagram.

3.6 Data Acquisition System

The data system is an IOTech 3000 that has a variety of input/output options. Setup and control of the data system uses the IOTech software. A view of the data system is provided in Figure 3.5. Temperature data is recorded during each deposition from two Type K thermocouples (TC) mounted on the outside edge of the baseplate centered along the length of the deposition. The TCs are placed to coincide with a measurement point in the corresponding DATM model. The TC dimensions are provided in Figure 3.6 and are not electrically shielded. Each thermocouple is connected to a converter that is used to transform the non-linear low magnitude (millivolt) voltage generated by the thermocouple to a standard 0-10V analog output signal that can be registered by the data system. Additionally, the thermocouple converter output and input range relationship is validated by implementing a 1:1 ratio between the input range (32 – 2192 °F) and the output range (0-10V) in the data system software. Then, a controlled temperature was applied to a thermocouple using a soldering iron, with a digital temperature readout, and the resulting 0-10V reading was confirmed, validating the TC signal conversion curve.

Equation 3.1 provides the conversion equation for Type K thermocouples with a temperature range of 32 – 932° F, and Table 3.5 provides the corresponding equation constants. The equations change depending on the temperature range of interest and the thermocouple voltage (V) must be compensated by the reference (cold junction) temperature reading:

$$T = c0 + c1 * V + c2 * V^2 + c3 * V^3 \ldots \ldots c9 * V^9. \tag{3.1}$$

Table 3.5: Type K Thermocouple Voltage Conversation Equation Constants for Temp Range of 32 – 932° F.

Constant	Value
c0	0

Constant	Value
c1	$2.508355*10^{1}$
c2	$7.860106*10^{-2}$
c3	$-2.503131*10^{-1}$
c4	$8.315270*10^{-2}$
c5	$-1.228034*10^{-2}$
c6	$9.804036*10^{-4}$
c7	$-4.413030*10^{-5}$
c8	$1.057734*10^{-6}$
c9	$-1.052755*10^{-6}$

The electrical interference that occurs during the high frequency start can overload a standard thermocouple card. Thus, transforming the signal from TC voltage to analog voltage prior to entering the data system ensures that the data system hardware is not damaged. Occasionally data are lost during this period but occurs before the weld is generated during a period where the data is not used.

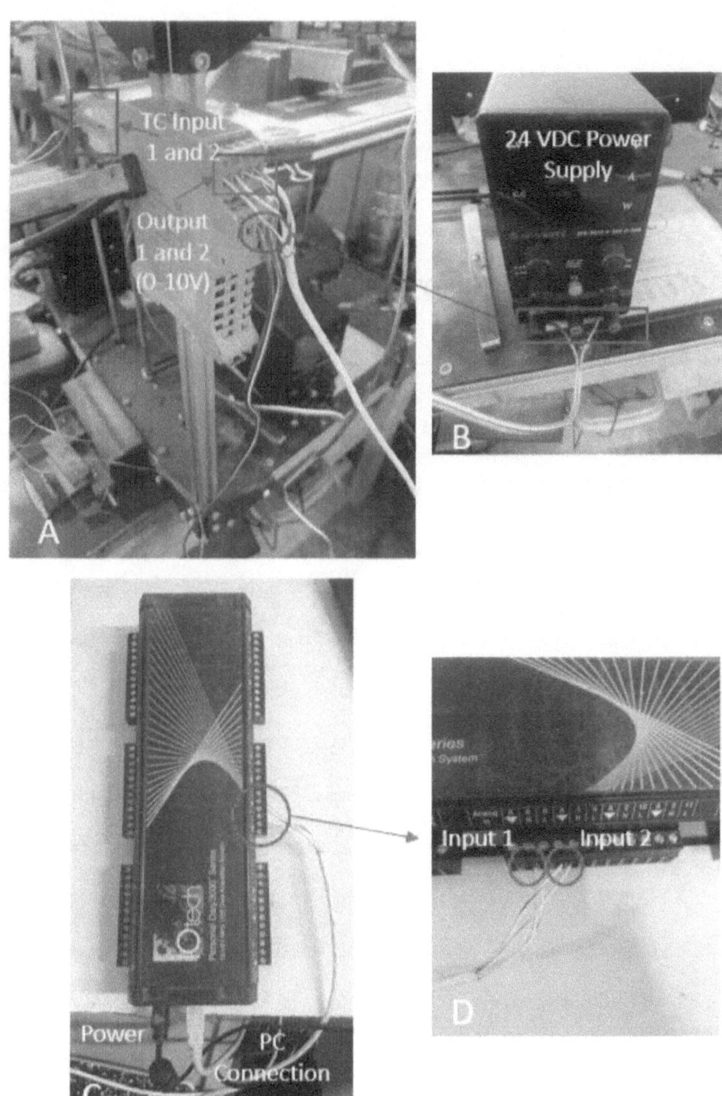

Figure 3.5: Data System Setup (A) TC signal to analog converter, (B) Analog Converter Power Supply, (C) Data Acquisition Module Power and PC Connections, (D) Data Acquisition Module Input Signal.

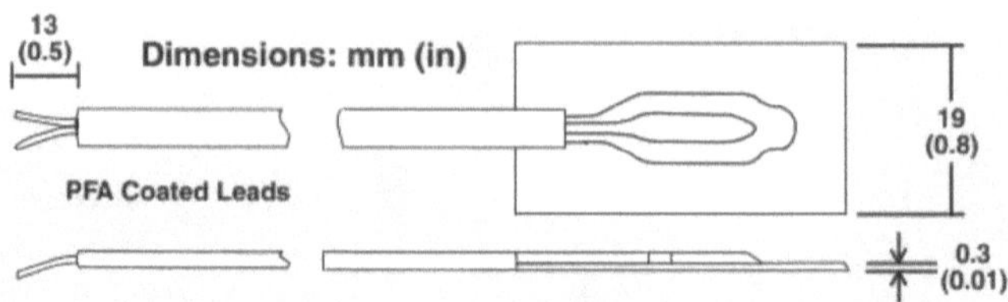

Figure 3.6: Thermocouple Dimensions.

3.7 Data Processing

There are two primary methods for post processing and viewing the data used in this study, including both exported DATM model data and recorded thermocouple data. Appendix A provides the MatLab code that was generated to view the Python data produced by the DATM model. The DATM model has the option to generate time history plots, which are saved as images instead of an inspectable and adjustable file format. To produce customizable plots with functionality to add axes, data sets, labels, etc. the MatLab code was generated. To start, the code reads in the Python data file and converts it to a MatLab array. This conversion requires a Python add-on for MatLab to get access to the functions required to interrogate the Python file. Once imported, the script uses a series of "for loops" to generate a plot for each data point that was generated in DATM. The final plot that is generated is an overlay of all node points on a single plot. In addition to plotting DATM results, the script has the functionality to overlay the DATM results with data recorded from thermocouples during a build. If this functionality is enabled, the user must select which node from DATM is going to be used for the comparison and which thermocouple channel from the data set is going to be used. The user defines the sample rate of the collected data, which is used to create a time vector of equal spacing. Finally, the user defines where in the recorded data time history the zero, or start point, should fall. This is used to align the start of the welding process with the start of the DATM model results.

Appendix B provides the MatLab code that is used to generate CCT plot overlays with the DATM data. The image in Figure 2.10 was imported into MatLab Image Viewer, which gave the ability to get XY coordinates for each pixel in the image. This information was used to create a map of pixel distance versus X axis length and was repeated for the Y axis. The result is an equation that converts the XY coordinate of any point on the plot into a physical unit (time and temperature) which can be used to compare directly against the time and temperature data from DATM. The provided code in Appendix B requires that Appendix A has already been executed. Using this script, the user is required to define where in the data set the start of the cooling curve of interest begins and ends (ModelTimeS, ModelTimeE). Additionally, the user defines which node from DATM is to be used for overlaying on the CCT diagram. The remainder of the code plots the CCT diagram lines and overlays the DATM data in a single figure. Multiple DATM traces can be plotted by using the "hold on" function in the plot where different start and end times can be used to generate multiple traces.

3.8 Test Specimen Nomenclature

Each test specimen that is generated receives a unique identifier, S##, where the ## placeholder is a non-repeating sequential number. Figure 3.7 shows the top-down (plan) view of a single pass deposition onto the base plate. After deposition is complete, the sample is sectioned into sub-samples shown as the segments in Figure 3.7. Each S## test specimen represents a single test sample, and the sub-sample nomenclature is shown in Figure 3.8.

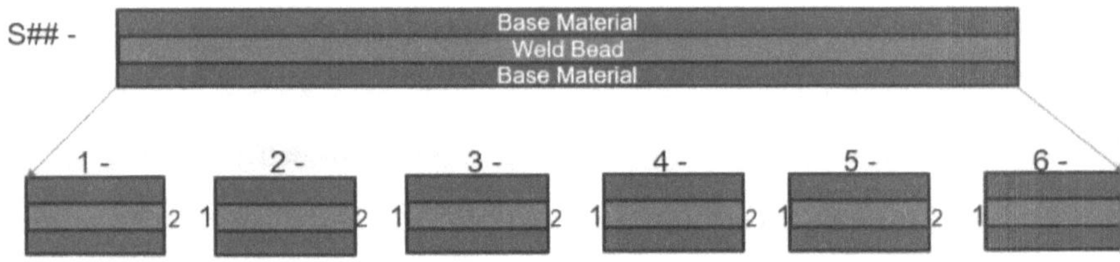

Figure 3.7: Plan view of the deposition Sample Nomenclature.

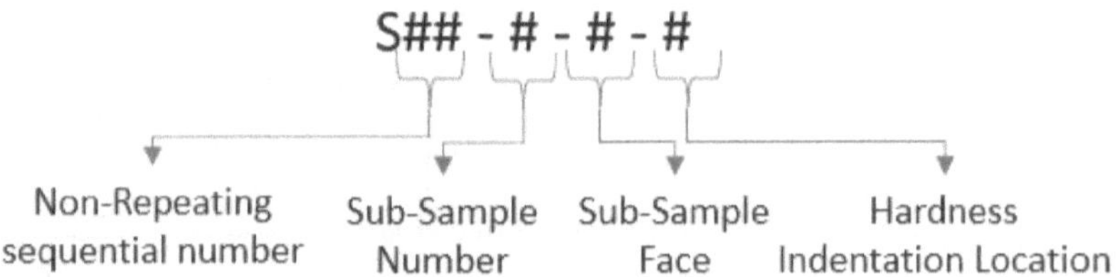

Figure 3.8: Test Sample Nomenclature.

3.9 Hardness Testing

For all hardness measurements, a Rockwell B scale was used on a Wilson Hardness Tester. Prior to each set of measurements, a calibrated sample was used to validate hardness tester functionality. For hardness testing, discrete points are taken along the build direction on the sub-samples. Figure 3.9 identifies the hardness locations across the build plate and along the GTAW deposition on the sub-sample face.

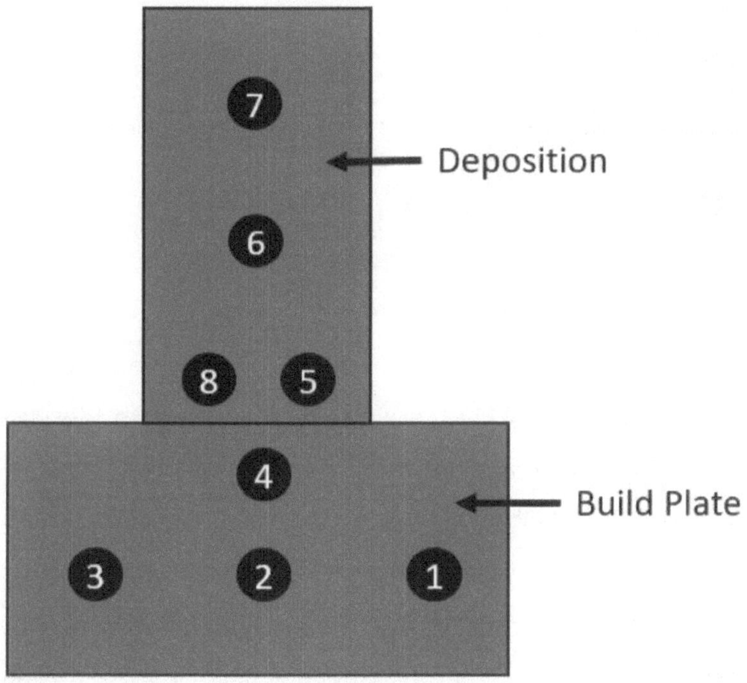

Figure 3.9: Hardness Indentation Locations for Sub-sample Faces.

As shown in the TTT diagram (Figure 2.9), the various phases can be identified by their hardness and are tabulated in Table 3.6 below. Test points 1 through 3 provide reference point consistency between samples for the base plate, which is not expected to change. Point 4 provides some insight into any microstructural changes in the heat-affected region of the weld. Points 5 through 7 provide hardness measurements along the height of the build. For a multi-layer build, those measurement points provide a comparison of the targeted heat treatment zone to the remainder of the build. For single layer builds, points 6 through 8 do not exist. For ten-layer builds, point 8 provides an additional measurement point adjacent to measurement point 5, which is in the targeted region for heat treatment.

Table 3.6: Tabulated Hardness Values for Various Fe-C Phases (Bhadeshia & Honeycombe, Steels: Microstructure and Properties, 2017) (Meyers & Chawla, 2009).

Phases	Rockwell	Vickers
Martensite	49-66 HRC	497-890
Bainite	24-49 HRC	252-497
Pearlite	100 HRB	252
Ferrite	47 HRB	-

Although ASTM standard E18-22 specifies spacing each indentation 3 diameters apart (or 0.19") and 2.5 diameters from the edge (0.16"), the geometry of the specimens in this study constrained abiding strictly by this. Violating this requirement is not typical for standard hardness testing, and can affect the measured values. On average, hardness measurements were 0.16" from the nearest edge but only 0.14" from the nearest adjacent indentation. Although not strictly within ASTM standards, this provided a qualitative assessment of heat treatment effectivity. This assumption was validated by machining the base material into the expected geometry of the deposition for a single layer. For this validation, all material is only the mild steel baseplate material, with no AM depositions present. Hardness values were recorded for comparison with indentations space equivalently to those taken on the deposition samples shown in Figure 3.9. An image of the as-machined hardness test validation sample is shown in Figure 3.10 and Figure 3.11 while the hardness results are given in Section 4.2.1.

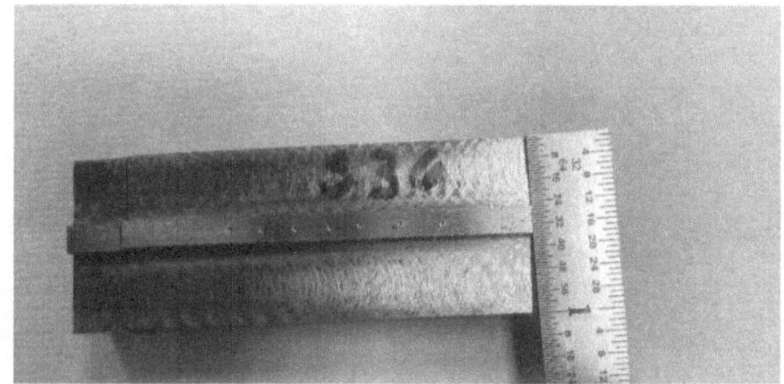

Figure 3.10: Plan view of the Machined Sample for Off-Nominal Edge Distance Hardness Testing.

Figure 3.11: Side View of the Machined Sample for Off-Nominal Edge Distance Hardness Testing of Sub-sample Faces.

3.10 Microscopy

After hardness testing, specimens were prepared for optical microscopy using a Zeiss XioVert.A1m Inverted Microscope. The sectioned samples are mounted in phenolic and metallographically prepared using standard procedures. The specimens were ground to remove the indentations and subsequent grinding marks using a series of grits from 180 to 1200. After each grit, the sample is washed with water, rinsed with acetone, rinsed with methanol, and dried. After grinding with the 1200 grit, the samples are polished using a range of polycrystalline diamond and aluminum oxide from 3μm to 1μm using Vel-cloth polishing pads. After polishing, macro photos are taken using a standard mounted digital camera with zoom lens. Next, the samples are imaged using the Zeiss microscope at 50X magnification with 5X objective. This provided information on the heat affected zone in addition to identifying any volumetric voids. After recording the low magnification images, the samples were etched to reveal the grain structure using Waterless Kallings. After etching, images were recorded over a range of magnifications to capture details of the deposited material along the build height.

3.11   Test Matrix

Table 3.7 summarizes the test matrix for this study. The single layer non-heat-treated sample provides a baseline reference that excludes the impacts of subsequent thermal cycling from the addition of subsequent layers. The single layer with heat treatment provides a baseline for weld parameters and heat treatment efficiency, and allows for accurate DATM to as-built comparison. The multi-layer non-heat-treated sample, similar to the single layer, provides a baseline for the properties of a multi-layer build without any impacts from heat treatment. Finally, the heat-treated multi-layer build combines the results of the previous samples and produces a multi-layer geometry with targeted heat treatment.

Table 3.7: Deposition Sample Matrix.

Test ID	Description	# of Passes	# of Heat-treated Layers	# of Hardness Samples	# of Microscopy Samples	Notes
1	Single Pass No Heat Treatment*	1	0	10	1	Baseline results for a single pass
2	Single Pass With Heat Treatment*	1	1	10	1	Baseline for heat treatment ability
3	Multi Pass No Heat Treatment*	10	0	6	1	Baseline for multi-pass with Temperature History
4	Multi Pass with Heat Treatment*	10	1	4	1	Heat treatment with temperature history

***The term "heat treatment" is used to reflect the *in-situ* process of applying heat to achieve altered microstructures**

Chapter 4.   Experimental Results

4.1 Modeled and As-Built Thermal Profiles

As discussed in Section 3.1, a thermocouple is placed on the build plate in a position that matches a predicted point in the numerical model. The comparison of measured thermal response to that of the model predictions for the baseplate is used to validate the model predictions for the deposition thermal profile. The following sections provide detail on the precise location of model prediction points and thermocouple placement coordinates. Additionally, measured versus predicted baseplate temperatures are provided along with deposition cooling curve CCT curve overlays. This data is provided for the single layer and 10 layer builds, both with and without *in-situ* heat treatments applied.

4.1.1 Single Layer Test Sample Thermal Profiles

Figure 4.1 and Figure 4.2 show the as deposited geometry for a single layer deposition and provides the origin of the coordinate system used to define the location of model thermal measurement points and build thermocouples. Table 4.1 provides the coordinates for each measurement point, Figure 4.3 provides a graphical view of the measurement points, and Figure 4.4 shows the model predicted thermal profile time history of the single layer deposition. The predicted thermal curve is compared to the CCT diagram from Figure 2.10 to determine the expected microstructure. Points 1, 2, 3, and 4 provide the base plate temperature along the Y=0 and Y=1 edge (with a 1-inch-wide base material), which correspond to thermocouple placement in the as-built samples. Points 5, 6, 7, and 8 provide temperatures along the length of the deposition at the base plate interface. Figure 4.5 provides a comparison of the predicted base plate thermal profile versus the as-built measured profile. Finally, Figure 4.6 provides the cooling curve of the predicted thermal profile overlayed on the CCT diagram.

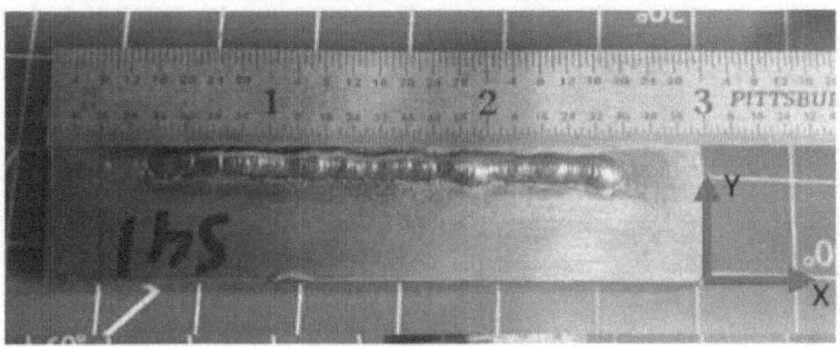

Figure 4.1: Single Layer Representative Deposition and Coordinate System (Plan View) – XY Intersection Represents the Origin.

Figure 4.2: Single Layer Representative Deposition and Coordinate System of Sub-sample Face (Side View) – XZ Intersection Represents the Origin.

Table 4.1: Single Layer Non Heat-Treated Model Measurement Points (inches) – Origin Defined in Figure 4.1 and Figure 4.2.

Measurement Point	X Coordinate	Y Coordinate	Z Coordinate
1*	-1.5	0	-0.25
2	-2	0	-0.25
3*	-1.5	0.98	-0.25
4	-2	0.98	-0.25
5	-1.5	0.5	0

Measurement Point	X Coordinate	Y Coordinate	Z Coordinate
6	-2	0.5	0
7	-0.5	0.5	0
8	-1	0.5	0
*** This measurement point corresponds to thermocouple placement on test samples**			

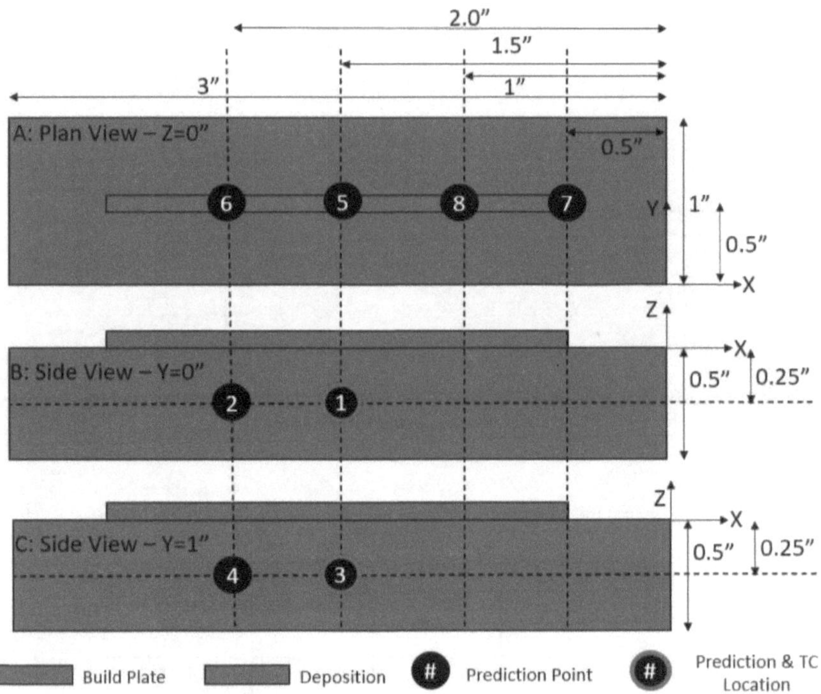

Figure 4.3: Single Layer Non Heat-Treated Model Measurement Points (inches) – Graphical View – A) Plan View B) Side View at Y=0" C) Side View at Y=1".

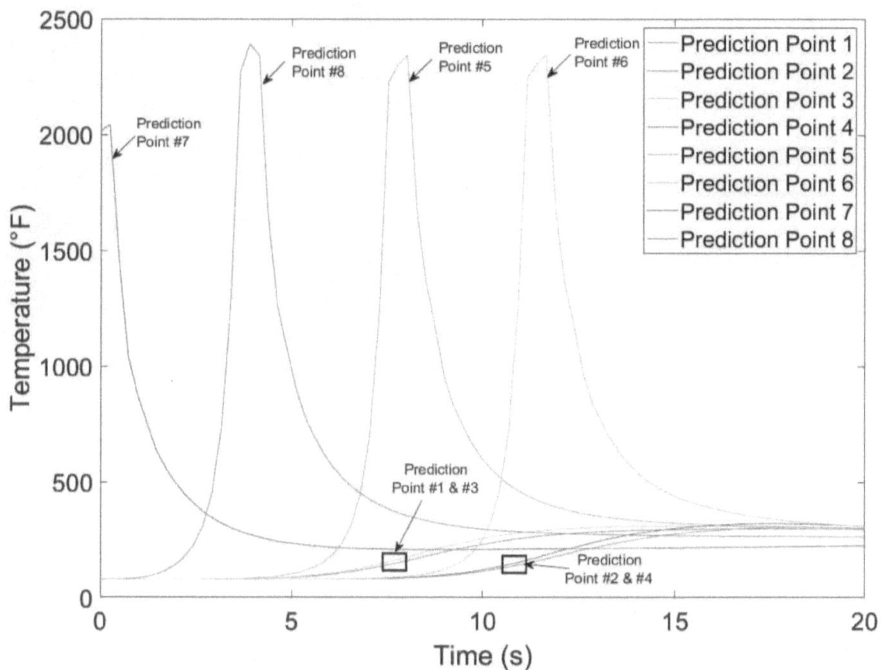

Figure 4.4: Single Layer No Heat Treatment Model Prediction Thermal Time History – S41 - Traces Correspond to the Measurement Points in Table 4.1.

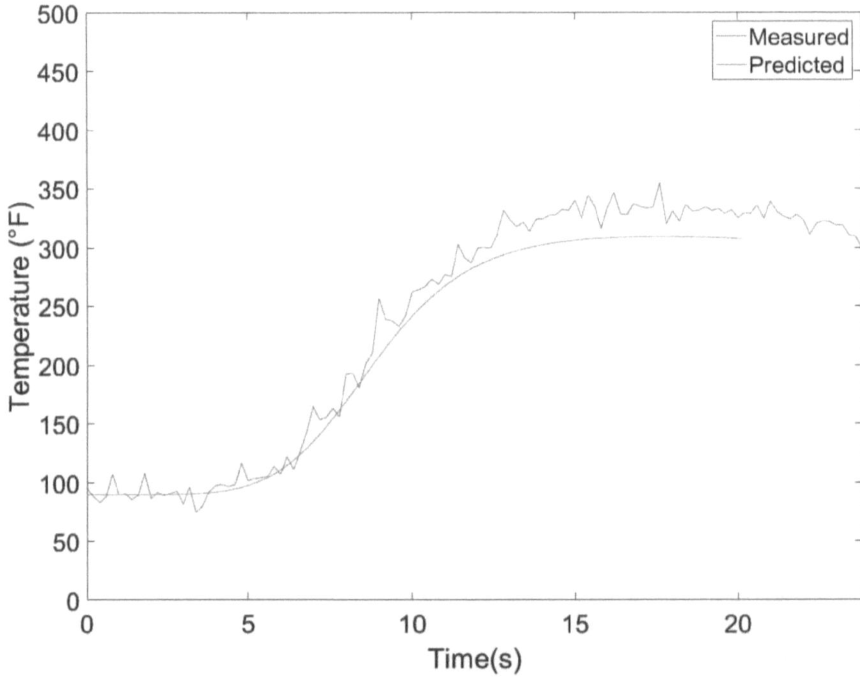

Figure 4.5: Single Layer No Heat Treatment Base Plate Thermal Profile Model vs. As-Built Comparison – S41 Measurement Point 1 from Table 4.1.

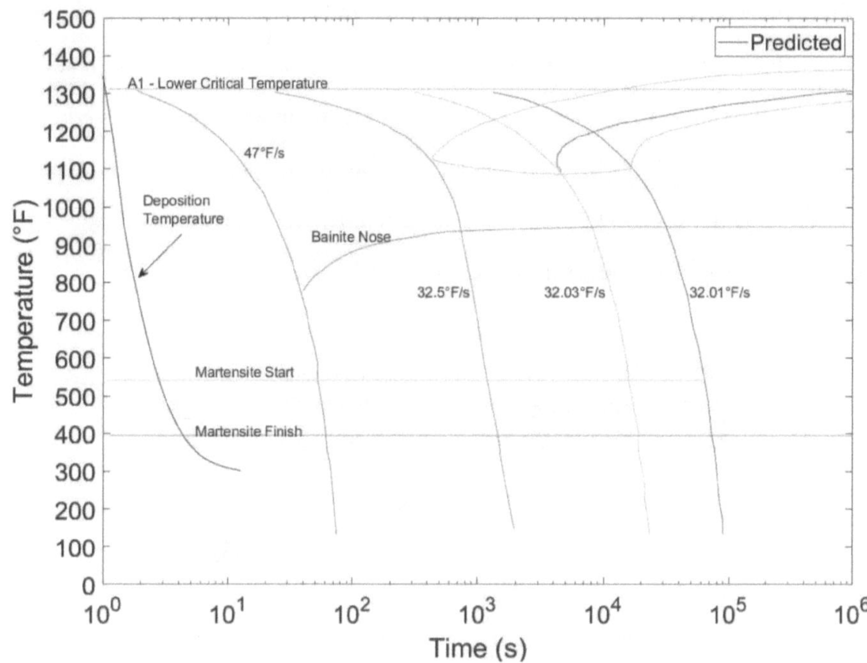

Figure 4.6: Single Layer No Heat Treatment Deposition Predicted Microstructure – S41 Cooling Curve with 4340 CCT Overlay.

The linear responses at the peaks of points 5 through 8 in Figure 4.4 are a function of the DATM model and data sampling and do not represent physical temperature responses. From Figure 4.5 it can be seen that the as-built base plate ramp rate aligns well with that of the model predictions and has peak values that align within 28 °F, showing acceptable deviation between model and build. Finally, **Figure 4.6** shows that for a single weld layer meeting the thermal profile of **Figure 4.4**, the primary microstructure is expected to be martensite. Figure 4.8, **Figure 4.9**, and Figure 4.10 provide the data for a single layer deposition with *in-situ* heat treatment applied. The *in-situ* heat treatment is applied over a one-inch span of the original single layer deposition centered on the deposition length. Heat-treatment consisted of two passes using

a current setting of 80 amps and a travel speed of 1.2 in/min. Figure 4.8 is the thermal time

history during the *in-situ* heat treatment, Figure 4.9 provides the model versus as-built base

profile comparison, and Figure 4.10 provides the deposition cooling profile overlayed on the

4340 CCT curve. Table 4.2 and Figure 4.7 provide a tabulated and graphical view of the

measurement points used for reporting temperatures.

Table 4.2: Single Layer Heat-Treated Model Measurement Points (inches).

Measurement Point	X Coordinate	Y Coordinate	Z Coordinate
1*	-1.25	0	-0.25
2	-1.875	0	-0.25
3	-1.875	0.98	-0.25
4	-3	0.98	-0.25
5	-0.75	0.5	0
*** This measurement point corresponds to thermocouple placement on test samples**			

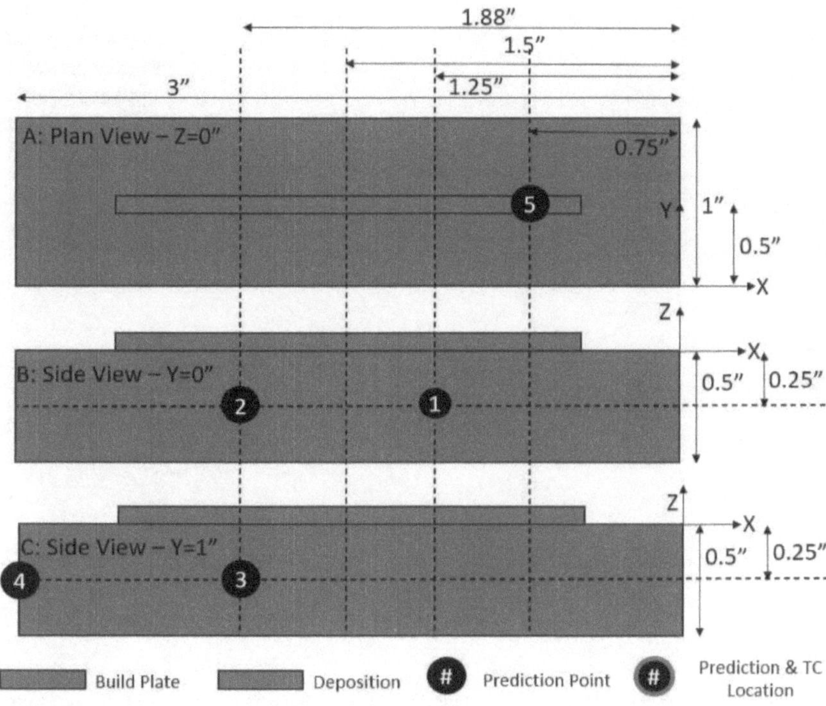

Figure 4.7: Single Layer Heat-Treated Model Measurement Points (inches) – Graphical view – A) Plan View B) Side View at Y=0" C) Side View at Y=1".

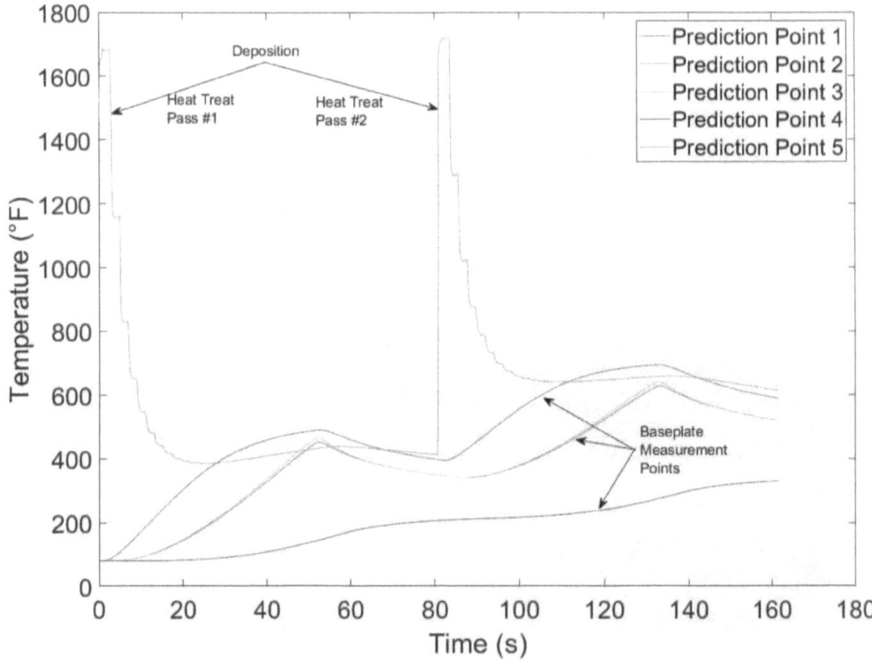

Figure 4.8: Single Layer with Heat Treatment Model Prediction Thermal Time History – S42 - Traces Correspond to the Measurement Points in Table 4.2.

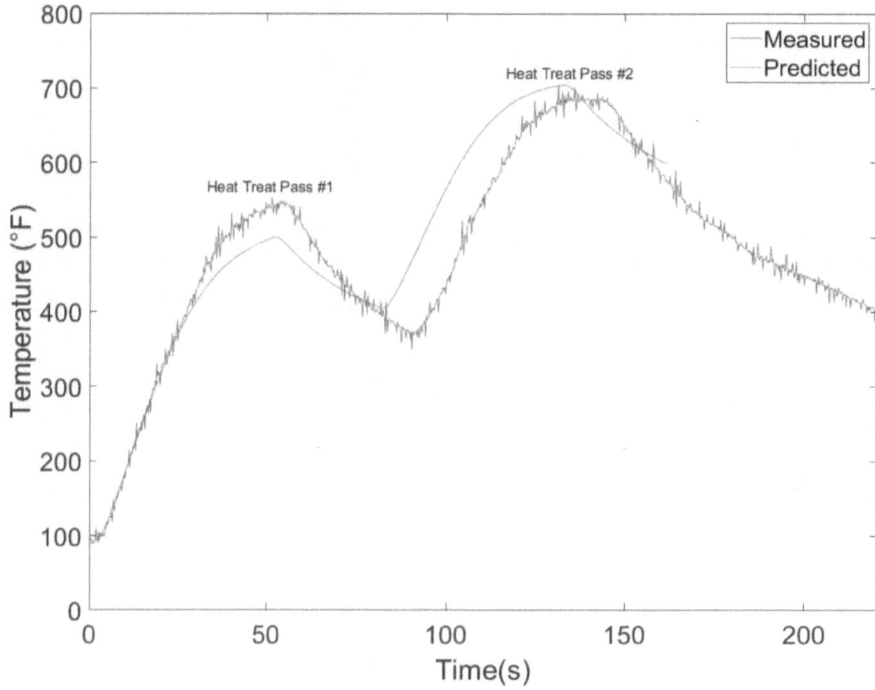

Figure 4.9: Single Layer with Heat Treatment Base Plate Thermal Profile Model vs. As-Built Comparison – S42 Measurement Point 1 from Table 4.2.

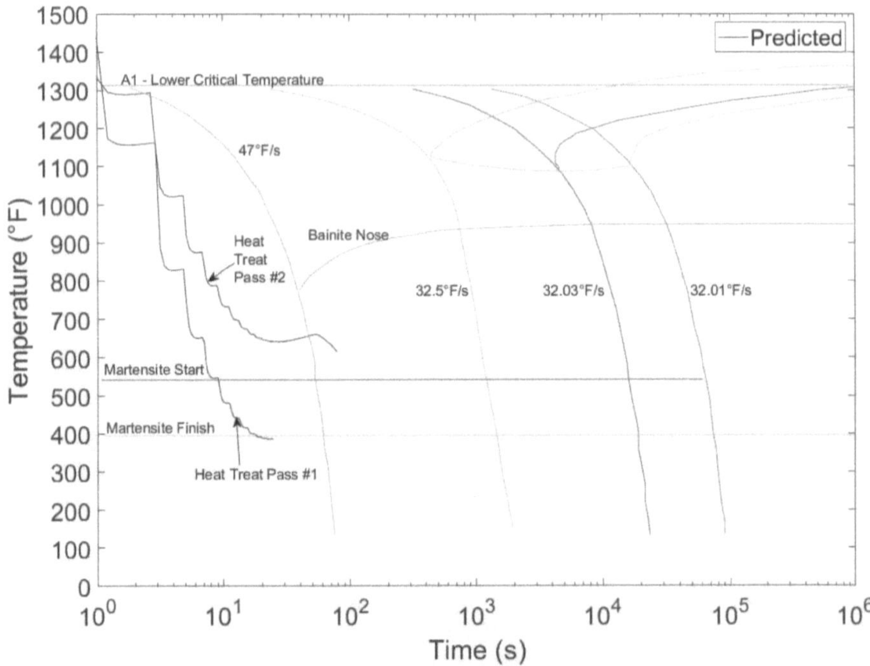

Figure 4.10: Single Layer With Heat Treatment Predicted Microstructure – S42 Cooling Curves with 4340 CCT Overlay.

Figure 4.8 shows an increase in the build plate temperature compared to the non-heat-treated sample shown in Figure 4.4. The increased temperature is due to the *in-situ* heat treatment exposing the base plate to elevated temperature for a longer duration. Figure 4.9 shows comparison between the prediction and as-built thermal data within approximately 24 °F, which is a negligible difference compared to the predicted peak temperatures of 1700 °F for the deposition shown in Figure 4.8. Finally, Figure 4.10 shows that the first pass of the *in-situ* heat treatment has a sufficient cooling rate for producing martensite. However, the slower cooling rate during the second heat-treatment pass sufficiently heats the surroundings resulting in a phase transformation to bainite, as targeted.

4.1.2   Multi Layer Test Sample Thermal Profiles

All multi-layer test samples are composed of ten layers. After each layer, the sample was rotated 180 degrees such that the starting point of the next layer occurred at the end point of the previous layer. Deposition current is decreased as the number of passes increases, as shown in Table 3.3, to maintain layer geometry throughout the build. Figure 4.11 shows representative as-built geometry for the nominal and heat-treated ten-layer samples, Table 4.3 provides the coordinates for the thermal measurement points, and Figure 4.12 provides a graphical view of the thermal measurement points. Figure 4.13, Figure 4.14, and Figure 4.15 provide the deposition thermal profile time history, comparison of predicted versus as-built base plate temperature profiles, and predicted cooling curves for each weld pass overlayed on the CCT curve, respectively. In Figure 4.13, four traces are shown but due to the symmetry in the measurement points relative to the heat input the traces overlap.

Figure 4.11: Ten-layer Nominal (A) and Heat-treated As-Built (B).

Table 4.3: Ten-Layer Non-Heat-Treated Model Measurement Points (inches).

Measurement Point	X Coordinate	Y Coordinate	Z Coordinate
1*	-1.0	0	-0.25
2*	-1.0	0.98	-0.25
3	-1.0	0.5	0
4	-1.0	0.5	0.196
*** This measurement point corresponds to thermocouple placement on test samples**			

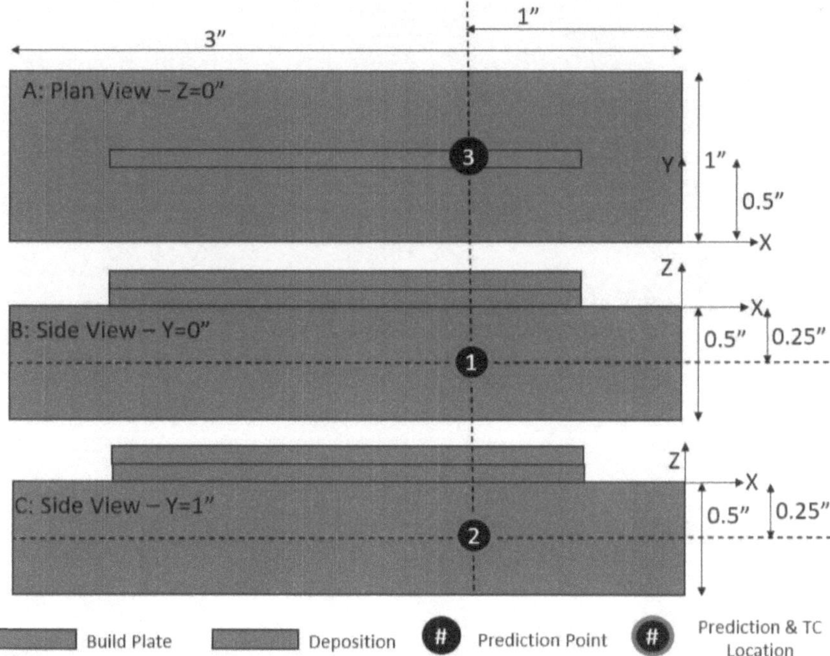

Figure 4.12: Ten-Layer Non-Heat-Treated Model Measurement Points (inches) – Graphical View – A) Plan View B) Side View at Y=0" C) Side View at Y=1".

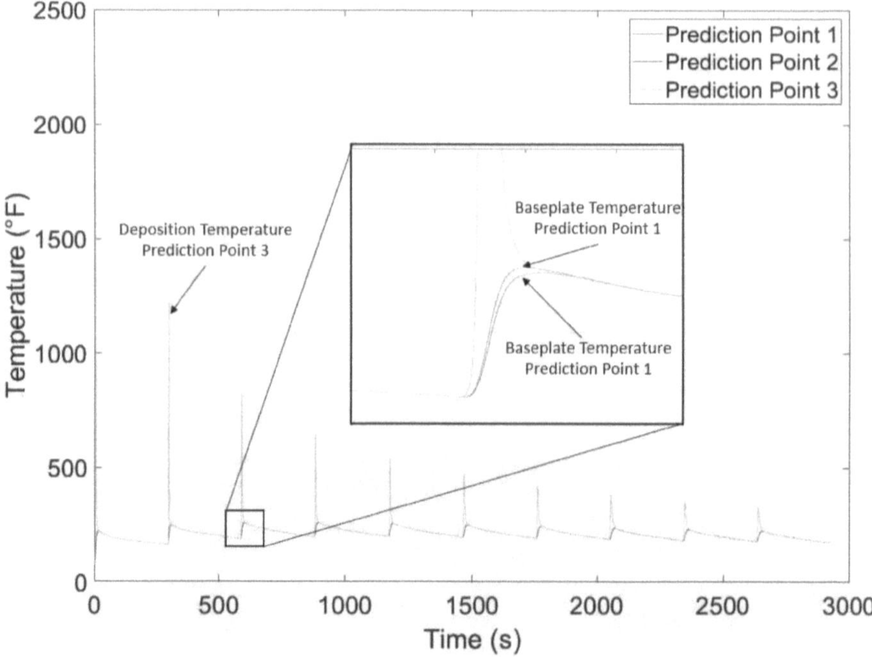

Figure 4.13: Ten-layer No Heat Treatment Model Prediction Thermal Time History – S43 – Traces Correspond to the Measurement Points in Table 4.3.

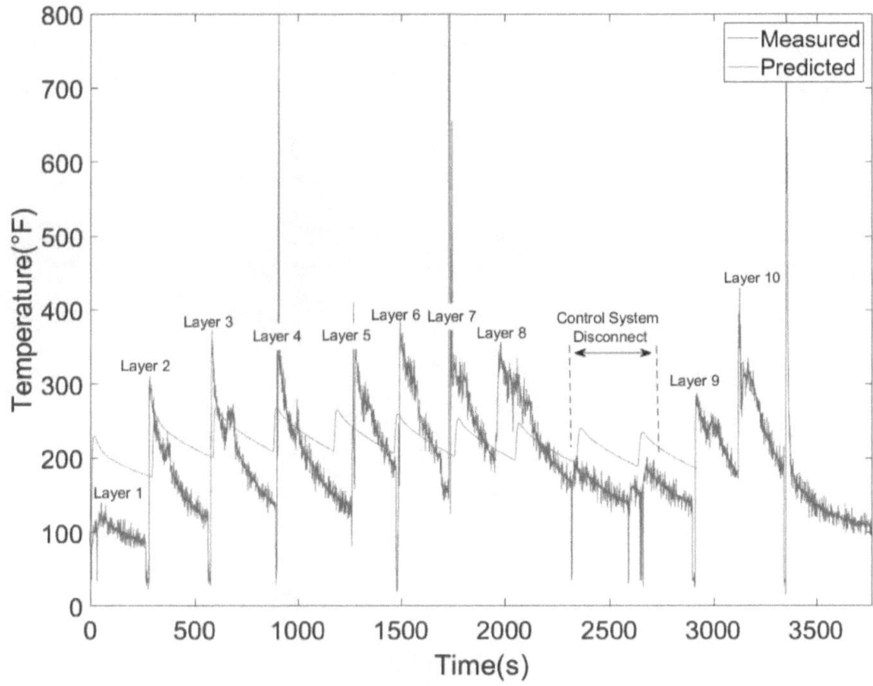

Figure 4.14: Ten-layer No Heat Treatment Base Plate Thermal Profile Model vs. As-Built Comparison – S43 Measurement Point 1 from Table 4.3.

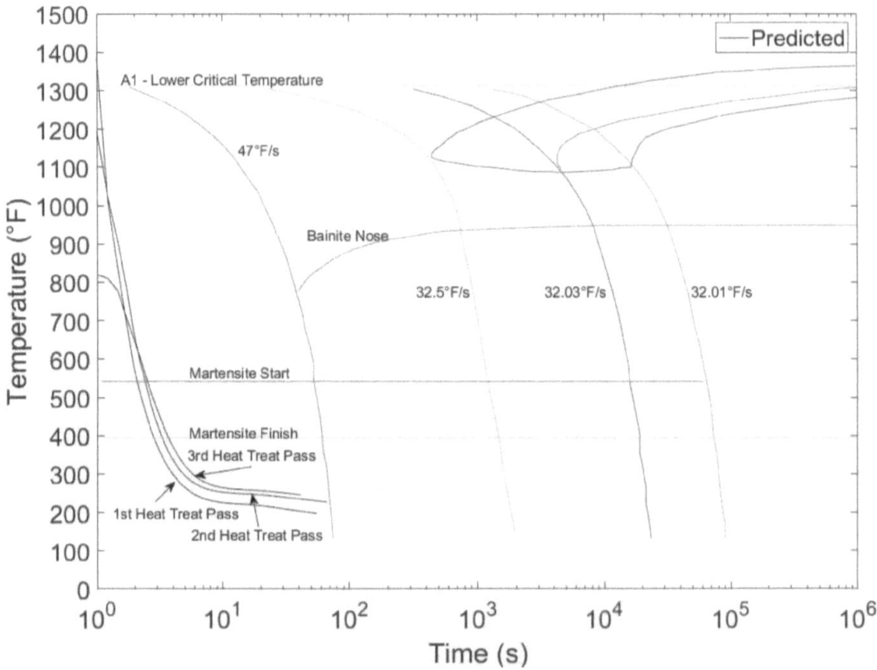

Figure 4.15: Ten-layer No Heat Treatment Deposition Predicted Microstructure – S43 Cooling Curves with 4340 CCT Overlay.

Measurement point 3 in Figure 4.13 indicates that after the second layer, the first layer of the deposition no longer exceeds the A_1 temperature. Thus, the 3rd layer becomes the cutoff for when *in-situ* heat treatment should be applied to the first layer without risk of annealing with subsequent deposition passes, giving a layer of margin for differences between model and as-built results. The abrupt spikes (positive or negative) in Figure 4.14 are due to the high electrical interference caused during arc initiation using the high frequency start functionality and the weld process in general. Additionally, at the initiation of layer eight there was a disconnect with the control PC that resulted in a longer delay than nominal between layer seven complete and layer eight starting. However, since this delay occurs after layer three, it does not have an impact on the first layers, which is the primary point of comparison to the *in-situ* heat-treated sample. Figure 4.15 confirms that after the second layer the temperature in the first deposition layer does not exceed the A_1 eutectoid temperature. Additionally, for both layers the predicted microstructure is primarily martensite. The ten-layer samples with *in-situ* heat treatment applied have thermal results that combine those of the non-heat-treated ten-layer samples, and the heat-treated single layer build. The heat treatment modelling results for the single layer build are applicable to the ten-layer build since the heat treatment occurs before the remaining layers (thermal mass) are added. Thus, no additional *in-situ* thermal modelling was required for the ten-layer build. Figure 4.16 provides the thermocouple placement on the build plate and Figure 4.17 shows the sequence of layer depositions versus the heat treatment pass location and targeted heat treatment region. Figure 4.18 shows the base plate thermal profile for a ten-layer build with *in-situ* heat treatment applied after layer three.

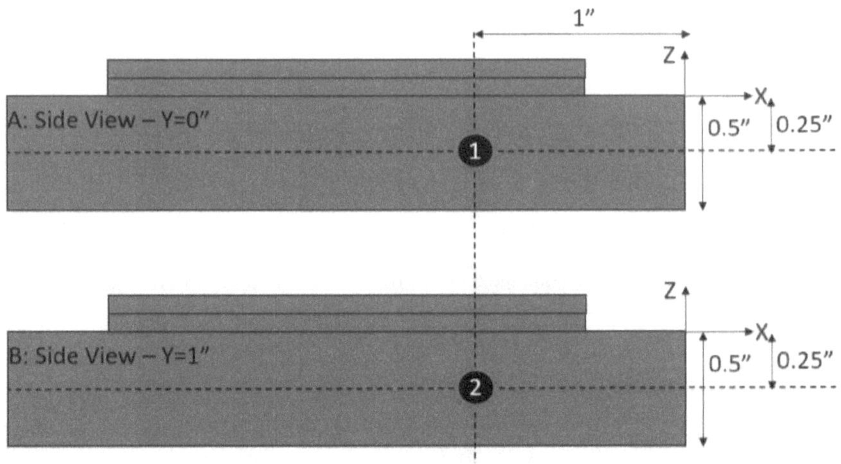

Figure 4.16: Ten Layer Heat-Treated Build Plate Thermocouple Placement (inches) – Graphical View – A) Side View at Y=0" B) Side View at Y=1".

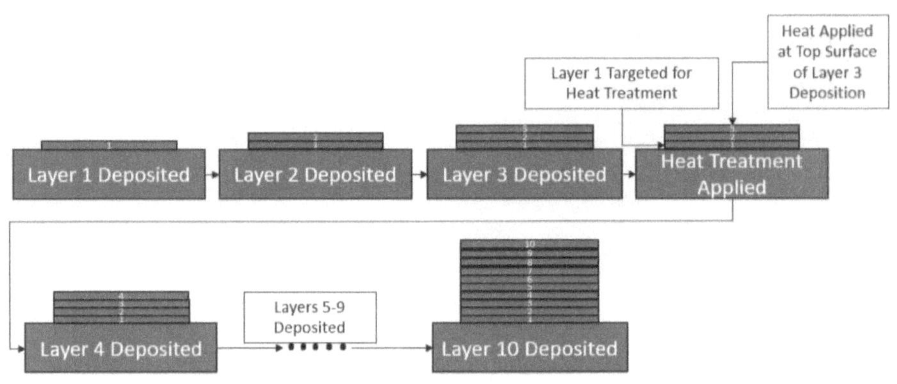

Figure 4.17: Ten Layer Heat-Treated Sample Deposition and Heat Treatment Sequence.

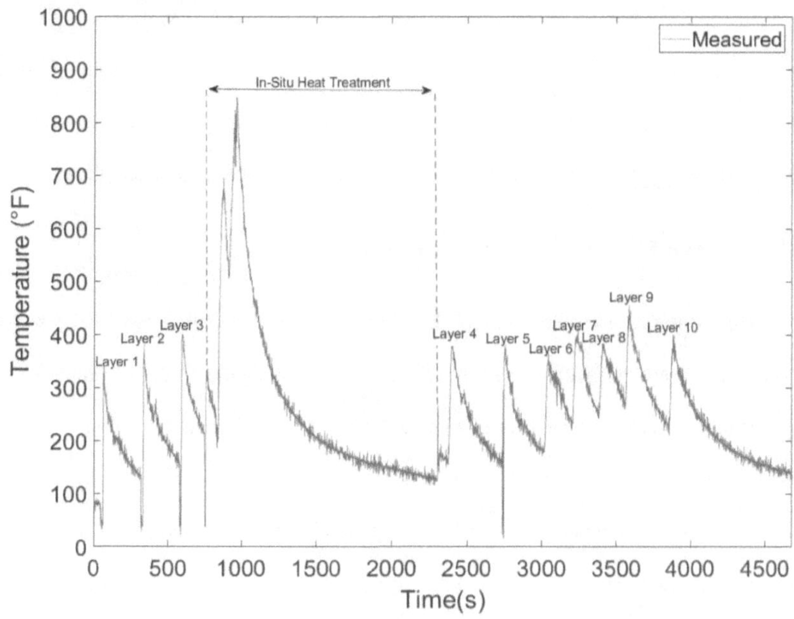

Figure 4.18: Ten-layer *In-situ* Heat Treatment Base Plate Thermal Time History – S45.

Figure 4.18 shows that during the heat treatment phase, the base plate temperature roughly doubles from 400 °F to 800 °F. The average baseplate temperature during layers one through ten (not including heat treatment) matches that of the ten-layer build without heat-treatment, as expected. The increased base plate temperature is desired, such that the cooling rate decreases, resulting in a change in phase in the heat-treated zone.

4.2 Hardness Test Results

As discussed in Section 3.9, the extracted subsamples were subjected to hardness testing. The hardness test results are used for comparison to the microstructure predictions of Section 4.1. Additionally, Chapter 3 introduced the discrepancy between the maximum edge distance

achievable with the as-built samples versus that of the associated ASTM standards. The
following sections provide qualitative validation of the hardness test results given the edge
distance discrepancy, followed by hardness measurement summaries for the single and 10-layer
samples. These results include both the heat-treated and non-heat-treated samples.

4.2.1 Hardness Measurement Validation Results

Hardness measurements were taken from the machined sample shown in Figure 3.10.
Measurements were taken along the base of the sample meeting the ASTM minimum spacing
requirements, then along the machined sample for comparison. Figure 4.21 provides a summary
of the hardness measurement results. Test points 1 through 7 were taken along the spine of the
machined section as shown in Figure 4.19. Test points 8 through 12 were taken along the base of
a sectioned sample as shown in Figure 4.20. Test points 13 through 16 were taken on the cross
section of the machined section as shown in Figure 4.20. For each measurement point taken
along the machined section, the distance to the nearest geometry (adjacent indentation or sample
edge) is provided in Table 4.4.

Figure 4.19: Hardness Validation Measurement Locations Along the plan view of the Machined Width (Plan View).

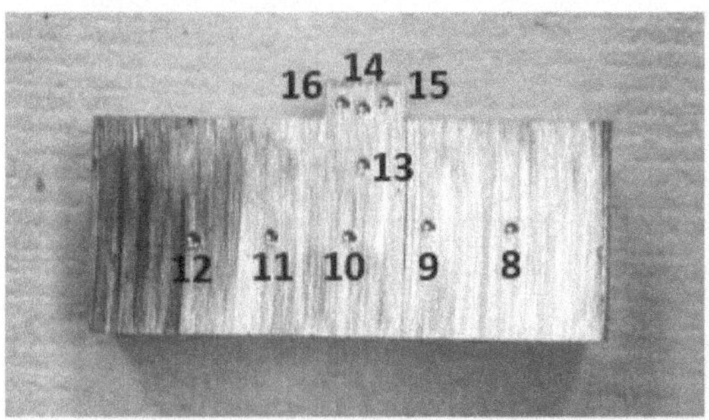

Figure 4.20: Hardness Validation Measurement Locations Along the Cross Section (Side View) Corresponding to Figure 3.9.

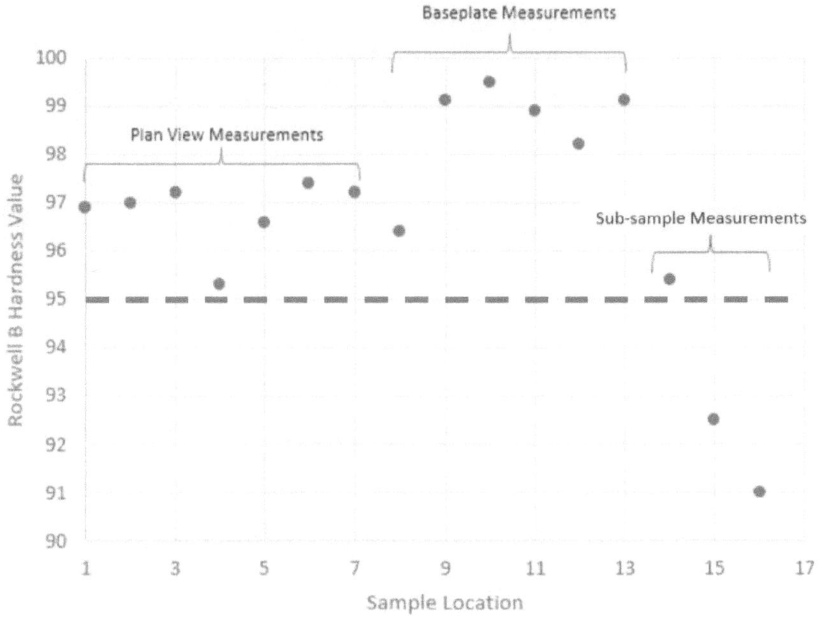

Figure 4.21: Hardness Measurement Methodology Validation Results – HRB Scale With Dashed Line Indicating Expected HRB 95 for Annealed 4340 from Table 3.2.

Table 4.4: Hardness Validation Edge Distance Summary.

Sample Location	Distance (in)
1	0.07
2	0.08
3	0.07
4	0.06
5	0.06
6	0.06
7	0.07
8-13	>0.19
14	0.06
15	0.04
16	0.04
Average	**0.06**

For determining impacts of not meeting the recommended edge spacing geometry, Figure 4.21 is used to compare the measurements taken on the machined section (1-7, 14) to the nominal baseplate measurements (8-13). Points 15 and 16 violate both the edge distance and the distance to adjacent indentation recommendations and have a notable reduction in hardness value as a result. The points taken on the machined section have an average HRB hardness of 96.6, while the base plate measurements have an average HRB hardness of 98.5. This results in approximately a 2% difference. In order to determine the applicability of this measurement approach to the current study, the percent difference due to edge distance violation is compared to the percent difference of the two primary phases of interest; martensite and bainite. From Figure 2.9 martensite and bainite have a percent difference of 36.5%. This indicates that the relatively low impact of taking HRB measurements on the deposition cross section without

meeting the required ASTM standard spacing is acceptable compared to the percent difference in

hardness measurements of the phases of interest in this study.

4.2.2   Test Sample Hardness Test Result Summary

Each test sample from the matrix shown in Table 3.7 was subjected to the hardness

testing as discussed in Section 3.9. Figure 4.22 shows an example of a single layer weld sample

after being hardness tested, while Figure 4.23 shows a ten-layer sample post hardness testing.

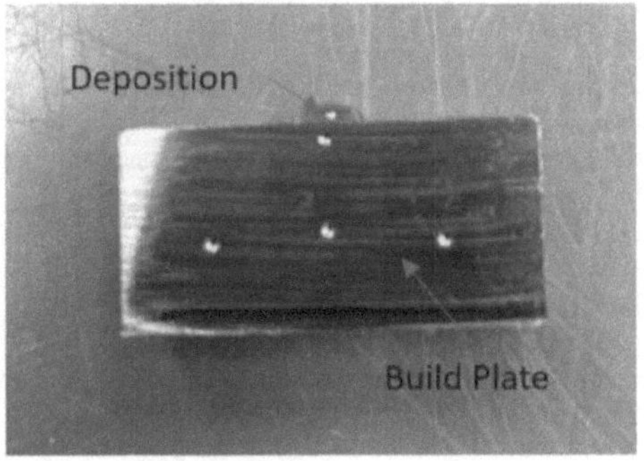

Figure 4.22: Single Layer Hardness Measurement Example.

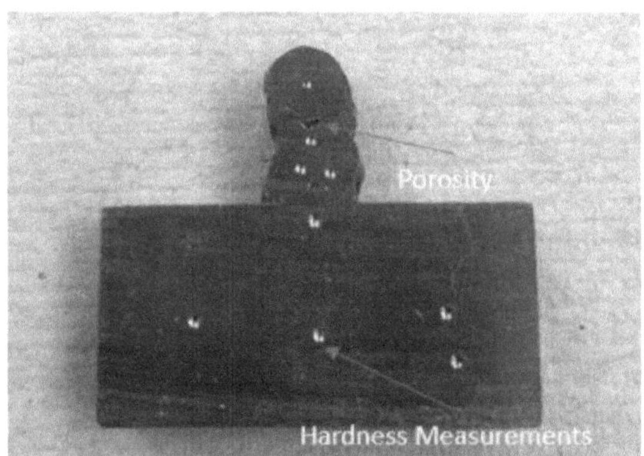

Figure 4.23: Ten-Layer Hardness Measurement Example.

Figure 4.24 shows the hardness results from the single layer test samples that were not exposed to *in-situ* heat treatment. The horizontal axis corresponds to the measurement points described in Figure 3.9 and Figure 4.22. Figure 4.24 shows the hardness results from the single layer test samples with *in-situ* heat treatment. Heat treatment targeted transforming the martensite that was present in the original sample to bainite, with reduced hardness.

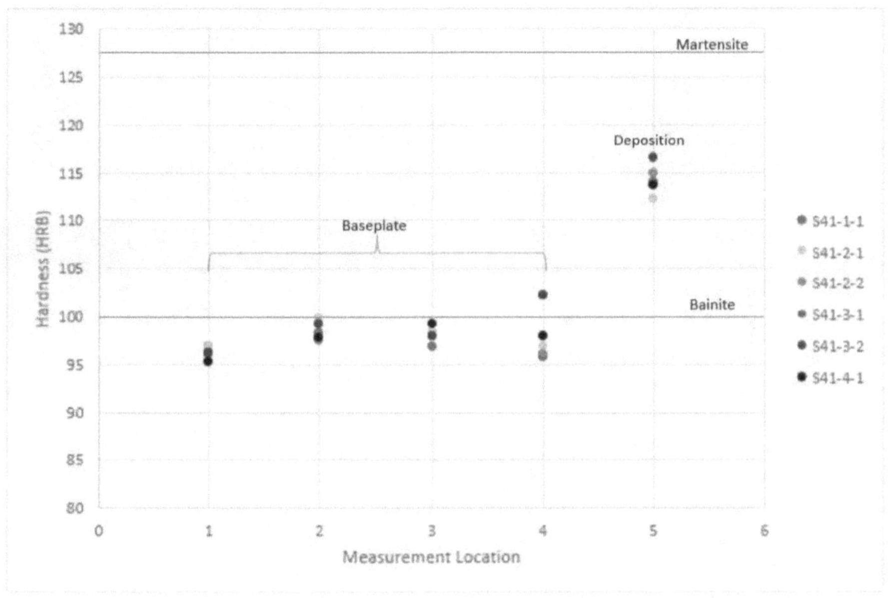

Figure 4.24: Single Layer Without *In-situ* Heat Treatment Hardness Results for the Sub-sample Face Measurements of Samples S41-1-1, S41-2-1, S41-2-2, S41-3-1, S41-3-2, S41-4-1.

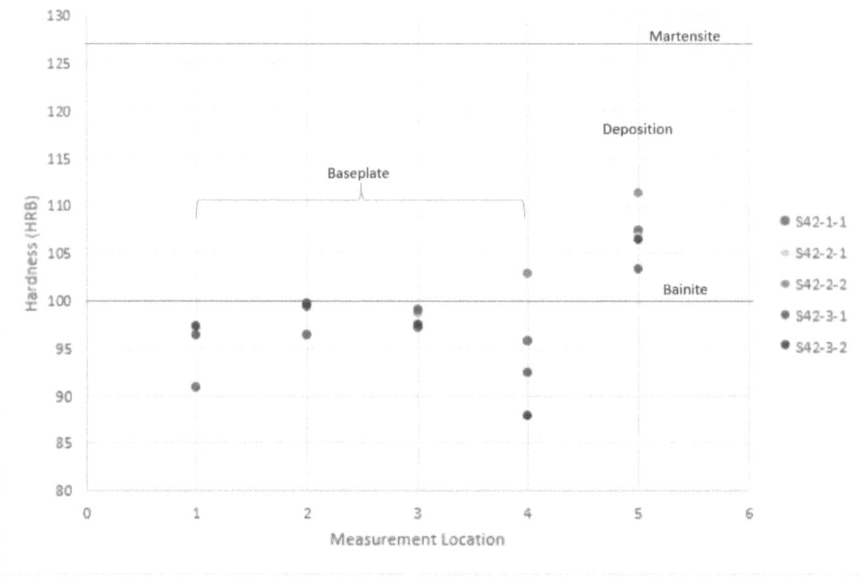

Figure 4.25: Single Layer With *In-situ* Heat Treatment Hardness Test Results for the Sub-sample Face Measurements of Samples S42-1-1, S42-2-1, S42-2-2, S42-3-1, S42-3-2.

To bound the range of martensite versus bainite present in a sample, an HRB of 102 is assumed as the minimum for bainite and a maximum of 119 for martensite (Bhadeshia & Honeycombe, 2017) (Meyers & Chawla, 2009) (ASM International, 1995). These values correspond to those provided in Table 3.6 and Figure 2.9. This range of hardness was used to linearly interpolate the relative percent composition of each phase using the rule of mixtures. Figure 4.24 shows that for measurement point 5, the average HRB value is 114, corresponding to a 71% martensitic composition. In contrast, Figure 4.25 shows an average HRB value of 106 at the same measurement point, corresponding to a 24% martensitic composition. These results are summarized in Table 4.5.

Figure 4.26 provides hardness results for the ten-layer weld samples that were not

subjected to *in-situ* heat treatment. Figure 4.27 shows the hardness results from an equivalent

ten-layer build sample that is subjected to *in-situ* heat treatment. The heat treatment targeted the

first layer of the build and was executed after the third layers deposition based on the results

shown in Figure 4.13. The heat treatment aimed to transform existing martensite in the first layer

to bainite, similar to the single layer builds.

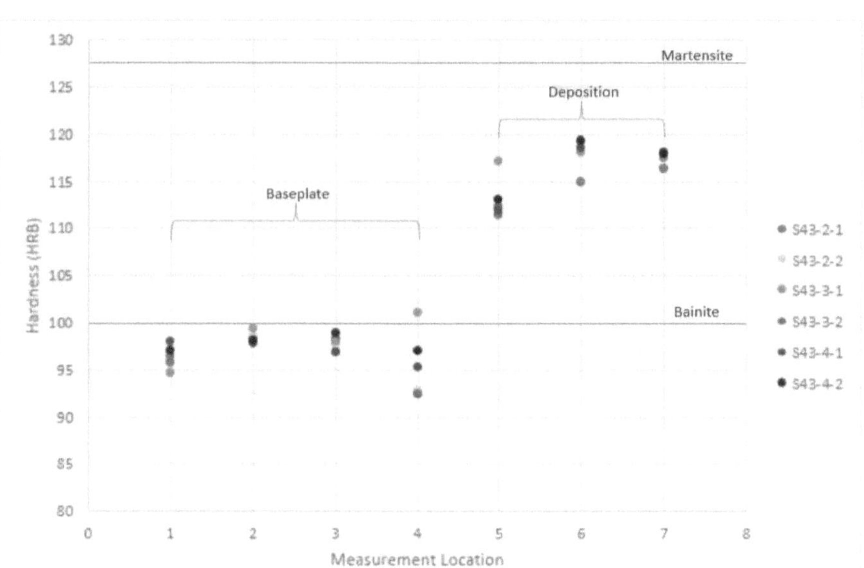

Figure 4.26: Ten-layer Without *In-situ* Heat Treatment Hardness Test Results for the Sub-sample Face
Measurements of Samples S43-2-1, S43-2-2, S43-3-1, S43-3-2, S43-4-1, S43-4-2.

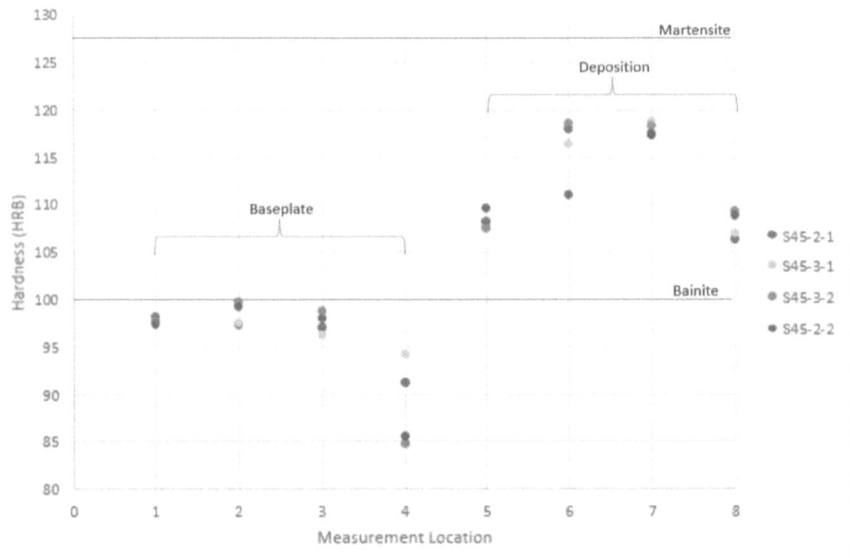

Figure 4.27: Ten-layer With *In-situ* Heat Treatment Hardness Test Results for the Sub-sample Face Measurements of Samples S45-2-1, S45-3-2, S45-3-2, S45-2-2.

The first layer of the build corresponds to measurement point 5. Figure 4.26 shows an average HRB value of 113 at this layer, which is analogous to the results seen at point 5 in Figure 4.24 for a single layer without heat treatment. Figure 4.27 shows an average HRB value of 107 at measurement point 5, which is comparable to the HRB value of 106 taken at the same measurement point in the single layer build shown in Figure 4.25. For the ten-layer non heat-treated sample measurement point 5 corresponds to a 68% martensitic composition, while the heat-treated samples of Figure 4.27 corresponds to a 29% martensite composition. Measurement point 4 in all samples has varying results, as exact placement of the indent is important due to the gradient present in the heat-treated region. Thus, the values are provided for completeness but are not used in this research. Additionally, measurement points 6 and 7 remain relatively

unaffected, as desired, resulting from the *in-situ* heat treatment being applied prior to their

deposition Table 4.5 provides a comparative summary of single and multi-layer build hardness

values with and without heat-treatment.

Table 4.5: Hardness Measurement Summary with Percent Composition – Single and Ten-Layer – With and Without Heat Treatment.

Measurement Source	HRB Value	% Martensite
Bainite	102	0%
Martensite	119	100%
Single Layer, Non Heat-treated, Point 5	114	71%
Single Layer, Heat-treated, Point 5	105	24%
Ten-layer, Non Heat-treated, Point 5	113	68%
Ten-layer, Heat-treated, Point 5	107	29%

4.3 Microscopy Results

Representative subsamples were prepared for optical microscopy as discussed in Section

3.10. The samples were mounted and polished after hardness testing. Images were taken in the

polished state to capture the as-deposited geometry. Afterwards, the samples were etched and

imaged and various magnifications at the points of interest. The following sections provided

details on the single and 10-layer builds for both the heat-treated and non-heat-treated conditions.

4.3.1 Single Layer Microscopy

Once polished and etched, a macro photo is taken to capture the state of the sample

during microscopy, and any visible inclusions in the sample. Figure 4.28 provides a representative image of a single layer sample etched with Waterless Kallings, where the heat affected zone and deposited layer are visible. A summary of images taken from the single layer samples are provided in Figure 4.29 and Figure 4.30, which includes un-etched and etched images and varying levels of magnification. Figure 4.29 represents sample S41 which is a non-heat-treated single layer deposition. In comparison, Figure 4.30 provides microscopy results for sample S42 that is a heat-treated single layer.

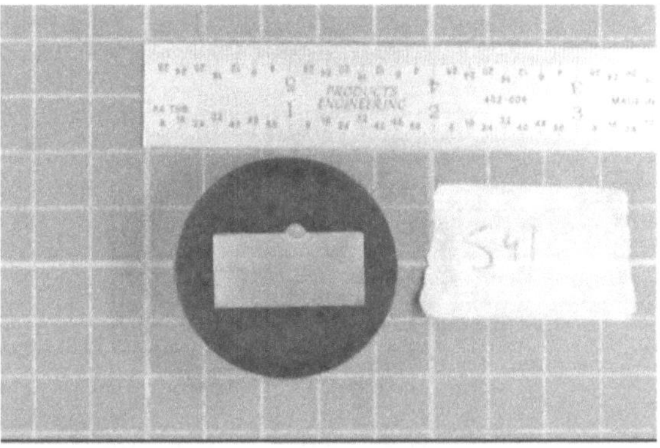

Figure 4.28: Representative Polished and Etched Sample of Sub-sample Face – S41 Single Layer Non-Heat-Treated Sample with Waterless Kallings Etchant.

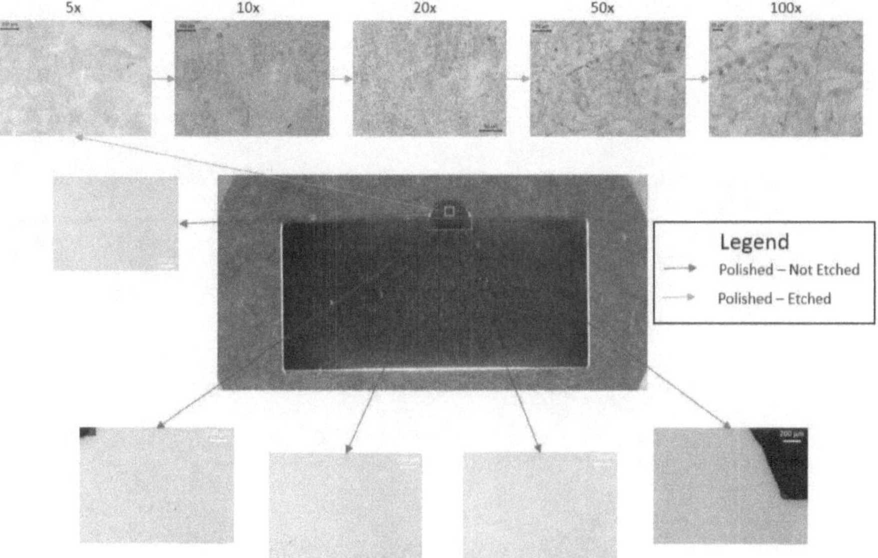

Figure 4.29: Single Layer Non-Heat-Treated Microscopy Results Summary of Sub-sample Face for S41. Etched Images Show a Range of Objectives from 5 to 100 X (50-1000X magnification). Samples That Were Only Polished are Shown at 50X Magnification.

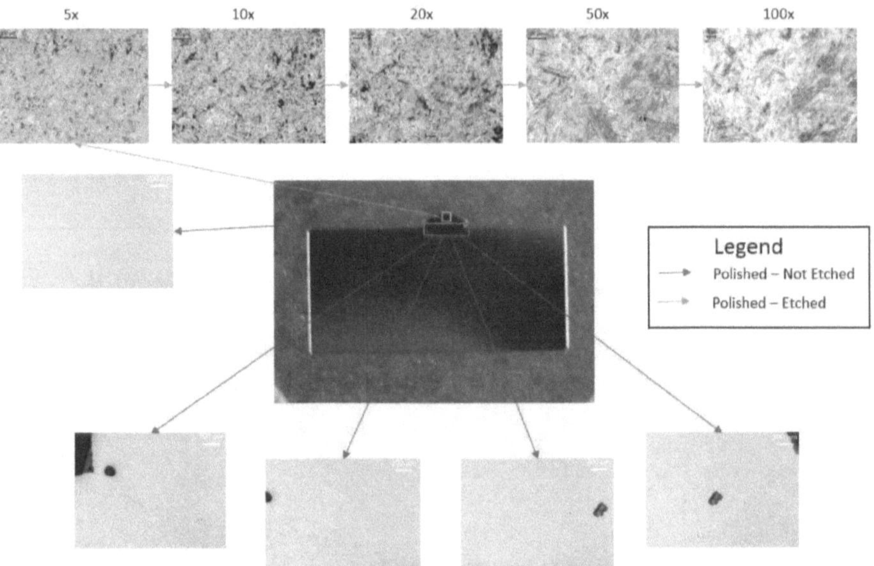

Figure 4.30: Single Layer Heat-Treated Microscopy Results Summary of Sub-sample Face for S42. Etched Images Show a Range of Objectives from 5 to 100 X (50-1000X Magnification). Samples That Were Only Polished are Shown at 50X Magnification.

Figure 4.31 shows the differing composition between the heat-treated and as-deposited (non-heat-treated) sample. Specifically, sample S41 contains large sections of the lightly colored plate martensite with smaller regions of darker colored bainite. Alternatively, sample S42 shows an increase in the presence of the bainite, and a reduction in the quantity and size of the martensitic plate. When compared to the hardness test results shown in Table 4.5, it is anticipated that heat-treated and non-heat-treated samples would consist of both bainite and martensite. The single layer microscopy results provide an independent verification of the hardness results.

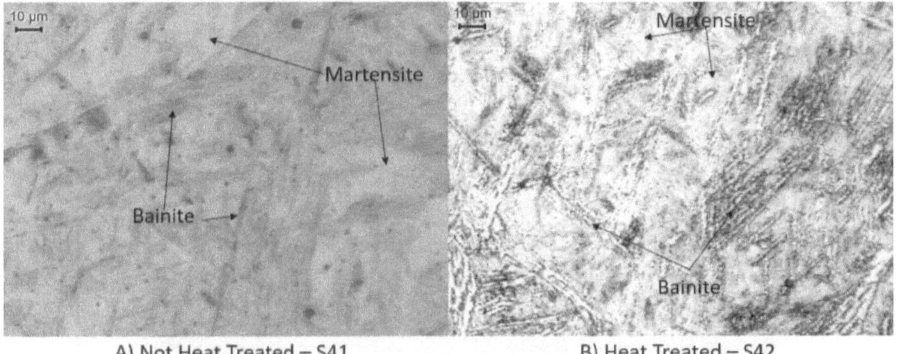

Figure 4.31: Single Layer Non-Heat-Treated sample S41 (A) and Heat-Treated Sample S42 (B).

4.3.2   Multi-Layer Microscopy Results

Similar to the single layer results of Section 4.3.1, Figure 4.32 shows a representative image of a multi-layer polished and etched sample used for microscopy. The heat affected zone, deposition layers, and inclusions are visible in the image. A summary of the ten-layer weld sample with no *in-situ* heat treatment is provided in Figure 4.33. Figure 4.34 provides the summary for a ten-layer weld sample with *in-situ* heat treatment performed after layer 3 in the deposition. Two comparisons are made between the *in-situ* heat-treated regions of the S45 sample. Figure 4.35 compares the heat-treated region of the S45 sample to layer ten of the same sample. Layer ten has considerably different boundary conditions and thermal time history compared to layer one of a ten-layer sample (with or without heat treatment). Figure 4.36 provides and additional comparison between the heat-treated region of the S45 sample and the same geometric region of the non-heat-treated S43 sample.

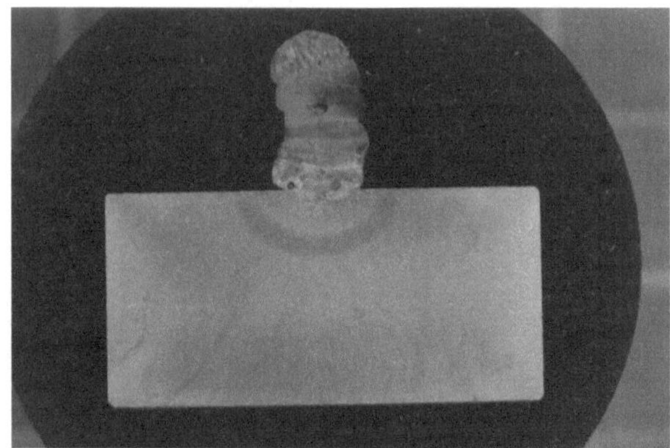

Figure 4.32: Representative Polished and Etched Sample of Sub-sample Face – S45 Ten-Layer Heat-Treated Sample with Waterless Kallings Etchant.

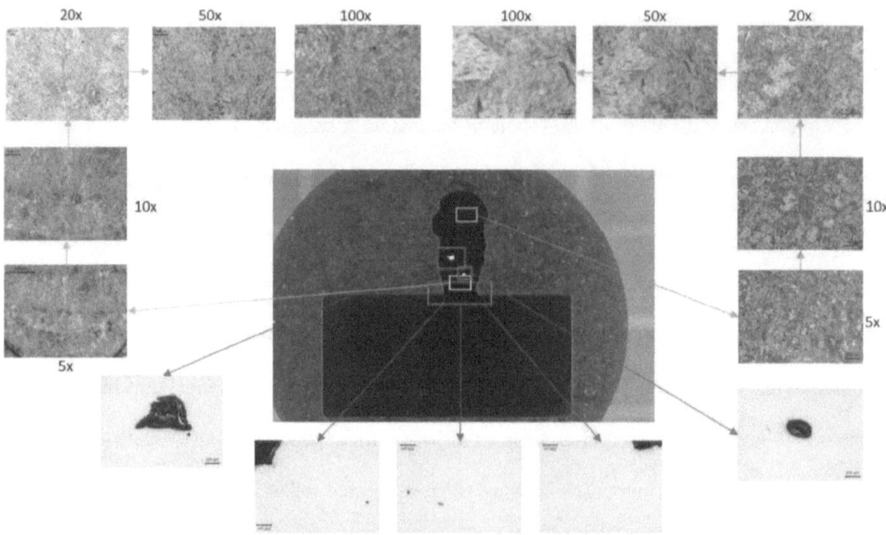

Figure 4.33: Ten-Layer Non-Heat-Treated Microscopy Results Summary of Sub-sample Face for S43. Etched Images Show a Range of Objectives from 5 to 100 (50 to 1000X Magnification). Samples That Were Only Polished are Shown at 50X Magnification.

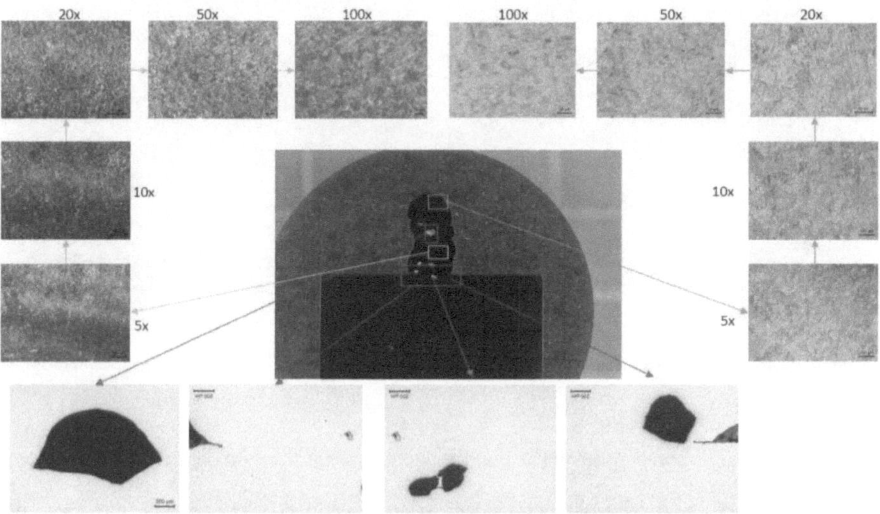

Figure 4.34: Ten-Layer Heat-Treated Microscopy Results Summary of Sub-sample Face for S45. Etched Images Show a Range of Objectives from 5 to 100X (50-1000X Magnification). Samples That Were Only Polished are Shown at 50X Magnification.

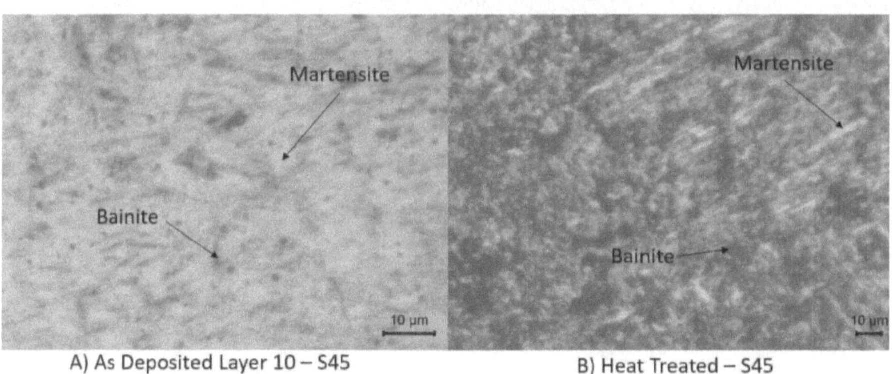

Figure 4.35: Ten-layer As-Deposited (A) and Heat-Treated (B) Layer 1 Microscopy Comparison for S45.

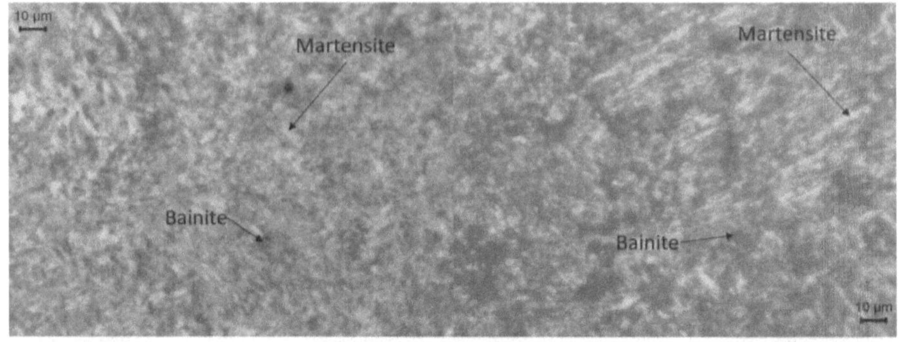

Figure 4.36: Ten-layer Non-Heat-Treated (S43) and Heat-Treated (S45) Layer 1 Microscopy Comparison.

Figure 4.35's comparison shows that the cooling rate at the tenth layer is still sufficient to produce a primarily martensitic structure, which is also supported by the results in measurement point 7 of Figure 4.27. The lighter colored martensite is more abundant than the darker colored bainite indicated in the figure, and the heat-treated region shows an increase in the bainite present. Similarly, Figure 4.36 once again shows an increase in the darker colored bainite when comparing the heat-treated region of S45 to the same region on the non-heat-treated S43. A final comparison is made in Figure 4.37 of a single layer heat-treated sample to a ten-layer heat-treated sample. Both images used for comparison show that bainite contributes a significant portion of the present structure, which is supported by the results shown in Table 4.5.

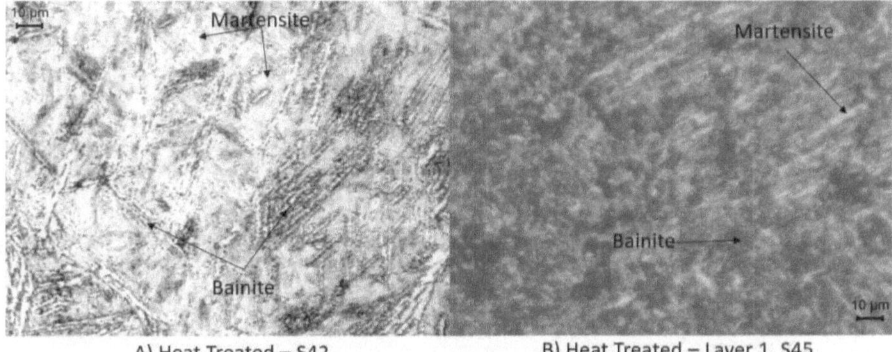

Figure 4.37: Heat-Treated Single Layer Sample S43 (A) and Heat-Treated Ten-layer Sample S45 (B) Microscopy Comparison.

Chapter 5.   Discussion

5.1 Thermal Modeling Versus As-Built

Single layer weld samples without heat treatment fit well within the capabilities of DATM, allowing for a direct comparison of as-built temperature results to model predictions. To account for heat treatment conditions, the model was used to simulate the weld conditions corresponding to heat treatment, even though those parameters would not produce a valid weld. That is, the model continues to add mass during heating, even though the prescribed heat conditions would not sufficiently melt the build material. Thus, DATM assumes a single layer is being welded at the lower current, or heat flux, conditions and it is assumed that the weld bead itself plays an insignificant role in the thermal profile, considering the low ratio of weld mass to base plate mass. The as-built weld sample thermocouple data is used to verify that the proper cooling rate was achieved to obtain the desired microstructure. The validity of this comparison is confirmed in Section 4.1.

For multi-layer welds, DATM does not lend itself well for directly modelling the heat treatment. Instead, the nominal results for a multi-layer simulation and the results from the single layer heat-treated weld bead are used to predict weld process settings. The standard multi-layer weld simulation provides insight into where in the layer sequence a particular point on the weld sample no longer exceeds the eutectoid temperature. This provides information on when in the build the point of interest could be heat-treated without being impacted by subsequent depositions. The heat treatment point of interest is within the first 3 layers of the build, such that the results of the single layer build with heat treatment can be used for determining heat treatment process parameters for the multi-layer build. The result is targeting a base plate temperature during the re-heat process that matches the base plate temperature during the single layer build. Similarly, the impacts of the additional weld material on the thermal profile are assumed to be negligible. If heat treatment were desired outside of the first three layers of the deposition, the thermal model software would need to be updated to allow for *in-situ* changes to weld parameters; heat flux and travel speed specifically.

The CCT plot overlays, such as Figure 4.15, do not necessarily accurately predict the combinations of phases present during initial deposition. However, they provide insight into what microstructural changes are expected during the *in-situ* heat treatment process. The curves do not claim that for the non-heat-treated condition that purely martensite exists in the structure. In fact, it is understood that due to the complex nature of the welding process and the considerable difference in dwell times and peak temperatures above the A_3 temperature, relative to traditional heat treatment processes, that other phases will be present during initial deposition. Thus, an *in-situ* heat treatment curve that claims to result in a bainite transformation indicates an increase in bainite and reduction in overall hardness of the test sample.

For the single layer weld samples, non-heat-treated and *in-situ* heat-treated, the predicted build plate thermal profile and the as measured thermal profile were within 25 °F as shown in Figure 4.5 and Figure 4.9. The *in-situ* heat-treated samples rely on precise matching of the modelled interpass time (*i.e.,* time between layers) to the time taken to reset the hardware and perform the second heat treatment cycle. However, Figure 4.9 shows that although the timing of the second heat treatment pass does not match perfectly with the model, the peak temperature of the model and test data match closely. Similarly, Figure 4.14 shows a disconnect in the timing of layers due to increased real world time to flip and prepare the sample and anomalies occurring at layer eight in the build. However, the peak base plate temperatures were in family with predictions and the hardness test results match well with the single layer predictions.

5.2 Hardness Testing

The Rockwell B hardness testing that was performed did not conform to ASTM standard practices, but it was shown through experimental results in Section 4.2.1 that the percent impact of not meeting edge distance requirements was insignificant compared to the predicted differences in the hardness of the phases of interest. Additionally, Table 4.4 shows an average edge distance measurement of 0.063" and the average edge distance of the test samples taken in the deposition region is 0.064" showing that the validation test is representative.

The results from Section 4.1 predict that the heat-treated samples should have an increased bainite content. Thus, the hardness measurements in the heat-treated region should be reduced compared to the non-heat-treated samples. Table 4.5 confirms the model predictions for increased bainite presence for both the single and ten-layer heat-treated samples. Additionally, the percent difference of hardness in the single layer non-heat-treated sample and ten-layer non-heat-treated sample at the heat-treated region is 0.9%. The percent difference of hardness in the single layer heat-treated and ten-layer heat-treated sample is 2%. Both results indicate process repeatability as the same phase transformation was targeted in each sample, regardless of layers present.

5.3 Microscopy

The intent of the microscopy in this study was to provide additional validation and verification of the hardness results and to confirm the predicted phase transformations occurred. Figure 4.31 confirms the presence of a higher bainite content in the single layer heat-treated sample, as noted by the increased cylindrical shaped bainite plates. Figure 4.15 shows that for a ten-layer sample without heat-treatment, the initial layers will not experience a phase transformation to bainite. The heat-treated region predictions fall within the heat treatment profile of the single layer build and is confirmed by both Figure 4.35 and Figure 4.36, which show an increase in the darker bainite present in the images. Although the microscopy provides a relatively qualitative assessment of heat treatment effectiveness, it provides evidence supporting the model predictions and heat treatment results.

5.4 Traditional GTAW Processing Versus GTAW DED MAM

A study of the impact of process parameters on the ultimate tensile strength of a welded joint determined the voltage and current settings of the GTAW process had the largest influence (Vijayan & Rao, 2018). Additional variables impacting weld or deposition quality are shielding gas selection, shielding gas flow rate, and speed. These same parameters must be considered when performing optimization of the GTAW DED process. The microstructure and mechanical properties of GTAW processes using 4130 steel have been studied in the as deposited and heat treated state (Souza Neto, Neves, Silva, Lima, & Abdalla, 2015). Additionally, the effect of the GTAW based weld process was compared to that of a laser deposition process determining that the laser process had a finer heat input area and thus had a faster cooling rate. This led to the laser process having finer grains that the GTAW based process. In the non-heat-treated condition both the laser and GTAW based process result in martensite formation. In contrast, with the application of traditional heat treatment methods the formation of ferrite and pearlite reduced the hardness of the weld samples and increased the ductility (Souza Neto, Neves, Silva, Lima, & Abdalla, 2015). These results are similar to the present research in that martensite formation is expected due to the rapid cooling rates resulting from the weld based process. In contrast, heat treatments in typical GTAW applications have utilized traditional long-duration heat treatments promoting formation of ferrite and cementite.

The GTAW based process has been studied as a method for increasing the hardness of the surface properties of 4340 steel parts as opposed to the use of carburizing, ion nitriding, chemical vapor deposition, laser cladding, or electron beam surfacing (Kumar, Ghosh, & Kumar, 2017). The decrease in heat input (weld current) resulted in increased cooling rates and the favored formation of martensite, while multi pass welds and higher heat input setting promoted bainite formation (Kumar, Ghosh, & Kumar, 2017). These results match closely with the microstructural results found in the present study and support the selection of process setting used for *in-situ* heat treatment. This is further supported by tensile testing of GTAW depositions at varying heat inputs and cooling rates that also indicated martensite formation with higher cooling rates and bainite formation with higher heat input or multiple pass welds (Kumar & Ghosh, 2018). Due to martensite formation associated with the rapid cooling rates of GTAW processing, traditional 4340 steel welding requires pre-heating and post-weld heat-treatment to reduce the potential for induced stress and cracking (Interlloy Engineering Steels and Alloys, 2011) .

Chapter 6.   Summary

The goal of this study was to explore the feasibility and use of *in-situ* heat treatments to spatially control microstructure evolution, thereby controlling the formation of site-specific microstructures. Additionally, the objective was to use the same processing equipment used for the AM processing to implement the *in-situ* heat treatment, which requires the use of an arc based AM process that de-couples heat input and feedstock. Numerical models were used to predict as-built microstructure, and to inform required process settings needed for the desired heat treatment effects. The as built samples were subjected to hardness testing and optical microscopy to verify that the desired microstructure had been achieved. Additionally, the study looked at two different geometries: single layer build and ten-layer build. The single layer build sought to simplify the thermal time history of the deposition, allowing for reduced variability and increased predictability of the resulting microstructure. The single layer build was also well suited for the existing thermal model (DATM) capabilities. The ten-layer build sought to confirm that in a multi-layer build the user could specify where in the build heat treatment would occur in order to achieve the same results as the single layer build. Optical microscopy was used to visually verify the phase transformations of interest had occurred, and hardness testing was used to quantify the impact of the achieved phase transformations on mechanical properties. This relationship is shown visually in Figure 6.1.

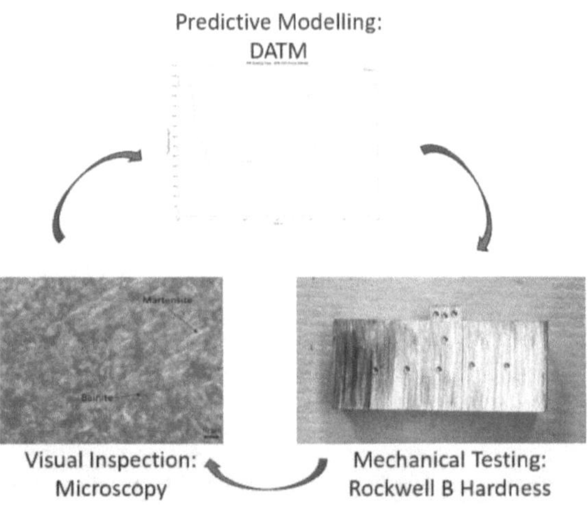

Figure 6.1: Relationship Between Model Results, Mechanical Testing, and Visual Inspection.

The starting point for the investigation was the single layer non-heat-treated sample. The DATM model was used to predict the cooling rate that would occur for this geometry and process settings, and the resulting CCT overlay is provided in Figure 4.6. Figure 4.6 shows that only martensite is predicted to form in the single layer build due to the high cooling rate of the sample. The hardness values provided in Figure 4.24 provide a basis of comparison for the remaining samples. Similarly, the microscopy results given in Figure 4.29 provide a baseline for non-heat-treated microstructure in this study. The first step in verifying *in-situ* heat treatment effectivity was to perform the heat treatment on the single layer sample. This also gave the basis for heat treatment procedure to be used for the multi-layer sample. The model predictions indicated that multiple passes would be required to heat a sufficient area of the sample to a temperature required for increasing the cooling rate for altering the phase decomposition. Thus, the model predictions for the single layer heat-treated sample are given in Figure 4.10. The predictions indicated for the slower cooling rate from the austenite, relative to those rates achieved during martensite formation, there would be some decomposition to bainite. This was confirmed by the hardness measurements for the single layer heat-treated sample, which are provided in Figure 4.25. Finally, visual inspection via optical microscopy was used to confirm the increase in the bainite phase was present. The microscopy summary is provided in Figure 4.30, and a comparison to the non-heat-treated sample is given in Figure 4.31. The combination of these results show that the predicted and desired microstructural changes were achieved. The ten-layer samples had the same objective with the increased thermal time history complexity and desire to heat treat only a portion of the sample. Model predictions for the non-heat-treated ten-layer sample are provided in Figure 4.15. The *in-situ* heat treatment model predictions for the ten-layer sample mimic those of the single layer (Figure 4.10) sample by assuming a negligible impact for the additional thermal mass of the first three deposition layers. This assumption would not be valid for larger build volumes, or for targeting a heat treatment region farther away from the build plate, where the thermal characteristics of the deposition drive the cooling rate instead of the base plate. Hardness results for the non-heat-treated ten-layer build are provided in Figure 4.26 and align with single layer non-heat-treated results with less than a 1% difference. Microscopy results for the ten-layer non-heat-treated sample are provided in Figure 4.33. Although there is additional bainite transformation due to heat cycles occurring without an *in-situ* heat treatment, there is commonality to the single layer microscopy results. Finally, the hardness results for the heat-treated ten-layer build are provided in Figure 4.27 and correlate to the heat-treated single layer with a less than 2% difference. Microscopy results for the heat-treated sample are provided in Figure 4.34, while Figure 4.35 and Figure 4.36 provide comparison to the non-heat-treated region of the alternate ten-layer build, and of the last deposited layer.

In summary, the model predictions for all four scenarios are validated by the hardness testing that has been performed and the visual inspection that was performed with optical microscopy. Where possible and applicable, further validation and verification is provided by comparing results between single and multi-layer build samples. This study has shown that thermal modeling can be used to predict and define process setting required to produce *in-situ* heat treatments. This was done by validating the predicted versus measured thermal profile of the base plate. Predicted microstructural changes were validated with hardness testing and microscopy. Furthermore, this study has validated the use of AM processes that decouple the heat source from the feed source as opportunities for performing *in-situ* heat treatments on AM parts. Finally, the study has confirmed the ability to achieve *in-situ* heat treatments at spatially defined locations within a build.

Chapter 7.   Future Work

7.1 Considerations for Modeling and Predictions of Present Phases

As discussed in Section 5.1, the current DATM model does not allow for modification of process parameters as a function of the build layer. Thus, there is no way to model the *in-situ* heat treatment at layers sufficiently far away from the build plate where the thermal response is no longer controlled by the build plate thermal response. Modification of the DATM program to allow for changing of travel speed, melt temperature, and pass time on a per layer basis would allow for the *in-situ* heat treatment used in this study to be sufficiently modelled. A simplified, but less capable, alternative would be to update the program to allow for customization of the build plate geometry. In which case, the build plate geometry could be constructed to replicate the geometry that would exist at the point where *in-situ* heat treatment would be applied. What is lost with this method is the thermal history of the build to that point. This requires the desired heat treatment to begin with the existing deposition at ambient conditions.

Section 2.3.4 presented that the AM process has a thermal profile that differs from traditional heat treatments. Specifically, the material may not complete the austenitic transformation during the heating phase due to the short duration above the Ac_1 or Ac_3 temperature. The quantity of austenite directly impacts the achievable decomposition during heat treatment. Future iterations of this work would benefit from incorporating approximations of austenitic decomposition to better understand the percent of phases that should be present after heat treatment. Incorporating the transformation approximation is especially applicable for this method of *in-situ* heat treatment since the heat source has a lower limit of temperature input capability (~2732 °F) that exceeds traditional heat treatment temperatures (~1652 °F). The temperature discrepancy is due to needing to maintain an arc from the tungsten electrode to the deposition and is shown in Figure 2.11.

7.2 Applicability to Other AM Processes and Heat Treatment Resolution

Section 2.1 provided an overview of various AM processes and their benefits relative to other AM processes. It is anticipated that other AM processes that incorporate a heat source that is independently controlled from the feed source will benefit from this study as the heat source separation is the key to in-situ heat treatment. Alternate heat sources could be used to perform the same function in AM processes without separated heat and feed sources, but doing so would lose the benefit of the in-situ heat treatment being executed with the same equipment used for the deposition. This is a benefit both monetarily and from a complexity reduction standpoint.

Figure 2.3 also shows that the achievable geometric resolution varies depending on the AM process being used. In general, the geometric resolution corresponds to the achievable resolution of the in-situ heat treatment. That is, a finer feed source typically has a finer heat source that can be used to heat treat with additional precision. Additionally, within a given AM process, parameters exist that would allow for increasing or decreasing the resolution. As an example, the size (diameter) and depth of the GTAW heat source can be varied by shielding gas used, tungsten polarity, and tip angle of the tungsten (Figure 7.1). Future advancement of this study would benefit from further quantifying the achievable resolution of *in-situ* heat treatment.

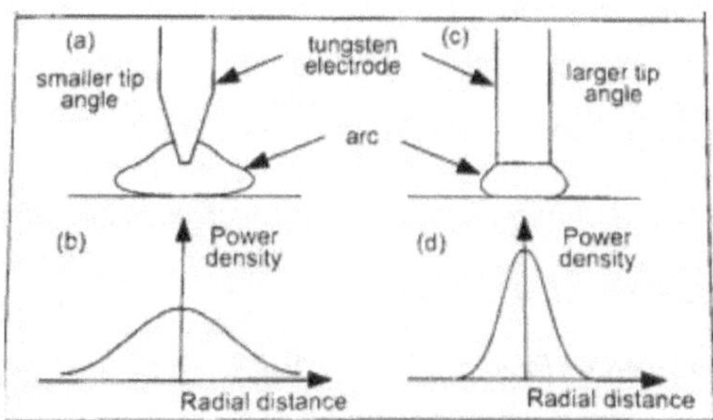

Figure 7.1: Effect of GTAW Electrode Tip Angle on Power Density (Kou, 2021).

References

ASM International. (1977). *Atlas of Isothermal Transformation and Cooling Transformation Diagrams.* Metals Park: ASM International.

ASM International. (1993). *ASM Handbook Volume 1: Properties and Selections: Itrons Steels and High Performance Alloys.* ASM International Handbook Committee.

ASM International. (1995). *Heat Treater's Guide: Practices and Procedures for Irons and Steels.* Materials Park: ASM International.

ASTM. (2012). Standard Terminology for Additive Manufacturing Technologies. *F2792-12A*, 1-3.

ASTM. (2016). Standard Guide for Directed Energy Deposition of Metals. *F3187.*

AZO Materials. (2012). *AISI 4340 Alloy Steel.* AZO Materials.

Baker, R. (1925). *United States Patent No. 1,533,300.*

Bearman, J., Bourell, D., Seepersad, C., & Kovar, D. (2020). Additive Manufacturing Review: Early Past to Current Practice. *Journal of Manufacturing Science and Engineering*, 142-152.

Beuth, J., Fox, J., Gockel, J., Montgomery, C., Yang, R., Qiao, H., . . . Klingbeil, N. (2013). Process Mapping for Qualification Across Multiple Direct Metal Additive Manufacturing Process. *24th International SFF Symposium*, (pp. 655-665).

Bhadeshia, H. (2015). *Bainite in Steels: Theory and Practice.* Boca Raton: Maney Publishing.

Bhadeshia, H., & Honeycombe, R. (2017). *Steels: Microstructure and Properties.* Cambridge: Elsevier.

Bhattacharya, S., Dinda, G., Dasgupta, A., & Mazumder, J. (2011). Microsturctrual Evolution of
AISI 4340 Steel During Direct Metal Deposition Process. *Materials Sciece and
Engineering*, 2309-2318.

Bourell, D. L. (2016). Perspectives on Additive Manufacturing. *Annual Review of Materials
Research*, 1-18.

Cai, X., Dong, B., Lin, S., Li, X., & Fan, C. (2022). Forming Characteristics and Mechanism of
Varialbe Polarity TIG-Based Wire Arc Additive Manufacturing of Al-Mg-Zn-Cu Alloy.
International Journal of Advanced Manufacturing Technology, 3007-3020.

Cerda, F., Goulas, C., Sabirov, I., Papaefthymiou, S., Monsalve, A., & Petrov, R. (2016).
Microstructure, Texture, and Mechanical Properties in a Low Carbon Steel After
Ultrafast Heating. *Matierals Science and Engineering*, 108-120.

Chen, C., Du, W., Zhang, H., & Zhao, X. (2023). Improvement of Microstructure and Mechanial
Properties of Stainless Steel TIG Based Wire Arc Additive Manufacturing by Using
AC/DC Mixed Current Waveform. *Journal of Materials Research and Technology*, 4355-
4366.

Costello, S., Cunningham, C., Xu, F., Shokrani, A., Dhokia, V., & Newman, S. (2023). The
State-of-the-Art of Wire Arc Directed Energy Deposition (WA-DED) as an Additive
Manufacturing Process for Large Metallic Component Manufacturing. *International
Journal of Computer Integrated Manufacturing*, 469-510.

Deng, Y., Li, Y., Di, H., & Misra, R. (2019). Effect of Heating Rate During Continuous
Annealing on Microstructure and Mechanical Properties of High-Strength Dual-Phase
Steels. *Materials Engineering and Performance*, 4556-4564.

Francois, M., Sun, A., King, W., Henson, N., Tourret, D., Bronkhorst, C., . . . Walton, O. (2017).
Modeling of Additive Manufacturing Processes for Metals: Challenges and Oportunities.
Current Opinion in Solid State and Materials Science, 198-206.

Frazier, W. E. (2014). Metal Additive Manufacturing: A Review. *Journal of Materials
Engineering and Performance*, 1917-1928.

Garcia-Colomo, A., Wood, D., Martina, F., & Williams, S. (2020). A Comparison Framework to
Support the Selection of the Best Additive Manufacturing Process for Specific Aerospace
Applications. *International Journal of Rapid Manufacturing*, 194-211.

Geng, H., Li, J., Xiong, J., Lin, X., & Zhang, F. (2017). Optimization of Wire Feed for GTAW
Based Additive Manufacturing. *Materials Processing Technology*, 40-47.

Gockel, J. D., & Beuth, J. L. (2013). Understanding Ti-6Al-4V Microstructure Control in
Additive Manufacturing via Process Maps. *24th International SFF Symposium*, (pp. 666-
674).

Gokhale, N. (2019). Experimental Investigation of TIG Welding Based Additive Manufacturing
Process For Improved Geometrical and Mechanical Properties. *Journal of Physics*.

Gokhale, N., Kala, P., Sharma, V., & Palla, M. (2020). Effect of Deposition Orientation on
Dimension and Mechanical Properties of the Thin-Walled Structure Fabricated by
Tungsten Inert Gas (TIG) Welding-Based Additive Manufacturing Process. *Mechanical
Science and Technology*, 701-709.

Gradl, P., Tinker, D., Park, A., Mireles, O., Garcia, M., Wilkerson, R., & McKinney, C. (2021).
Robust Metal Additive Manufacturing Process Selection and Development for Aerospace
Components. *Journal of Meterials Engineering and Performance*, 6013-6044.

Hidalgo, J., & Santofimia, M. (2016). Efect of Prior Austenite Grain Size Refinement by Thermal Cycling on the Microstructrual Features of As-Quenched Lath Martenstie. *Mettalurgical and Materials Transactions*, 5288-5301.

Honeycombe, R. (1995). *Steels: Microstructure and Properties.* Wiley.

Hoyt, J., Asta, M., & Karma, A. (2003). Atomistic and Continuum Modeling of Dendritic Solidification. *Materials Science and Engineering*, 121-163.

Interlloy Engineering Steels and Alloys. (2011). *4340 High Tensile Steel.* Retrieved from Interlloy: https://www.interlloy.com.au/our-products/high-tensile-steels/4340-high-tensile-steel/#:~:text=It%20is%20preferred%20that%20welding,prior%20to%20hardening%20and%20tempering.

Jeon, J., Seo, N., Jung, J.-G., Kim, H.-S., Son, S., & Lee, S.-J. (2022). Prediction and Mechanism Explanation of Austenite Grain Growth During Reheating of Alloy Steel Using Explainable Artificial Inteligence. *Journal of Materials Reasearch and Technology*, 1408-1418.

Karma, A., & Tourret, D. (2016). Atomistic to Continuum Modeling of Solidification Microstructures. *Current Opinion in Solid State and Materials Science*, 25-36.

Kinney, C. C., Yi, I., Pytlewski, K. R., Khachaturyan, A. G., Kim, N. J., & Morris, J. W. (2017). The Microstructure of As-Quenched 12Mn Steel. *Acta Materialia*, 442-454.

Kou, S. (2021). *Welding Metallurgy Third Edition.* Hoboken: John Wiley & Sons, Inc. .

Kumar, A., Gautam, S. S., & Kumar, A. (2014). Heat Input and Joint Efficiency of Three Welding Processes TIG, MIG and FSW. *International Journal of Mechanical Engineering and Robotics Research*, 89-94.

Kumar, S., & Ghosh, P. (2018). TIG Arc Processing Improves Tensile and Fatigue Properties of Surface Modified AISI 4340 Steel. *International Journal of Fatigue*, 306-316.

Kumar, S., Ghosh, P., & Kumar, R. (2017). Surface Modification of AISI 4340 Steel by Multi-Pass TIG Arcing Process. *Journal and Matierlas Processing Tech.*, 394-406.

Lee, S. (2013). Predictive Model for Austenite Grain Growth During Reheating of Alloy Steels. *Iron and Steel Institute of Japan*, 1902-1904.

Li, Y., Su, C., & Zhu, J. (2022). Comprehensive Review of Wire Arc Additive Manufacturing: Hardware System, Physical Process, Monitoring, Property Characterization, Application, and Future Prospects. *Results in Engineering*, 1-17.

Li, Z., Liu, C., Xu, T., Ji, L., Wang, D., Lu, J., . . . Fan, H. (2019). Reducing Arc Heat Input and Obtaining Equiaxed Grains by Hot-Wire Method During Arc Additive Manufacturing Titanium Alloy. *Materials Science and Engineering*, 287-294.

Lim, H., Abdeljawad, F., Owen, S., Hanks, B., Foulk, J., & Battaile, C. (2016). Incorporating Physicaly-Based Microstructures in Materials Modeling: Bridging Phase Field and Crystal Platicity Frameworks. *Modelling and Simulation in Materials Science and Engineering*.

Liu, H., Feng, T., Chen, C., & Chen, H. (2023). Study on the Relationship Between Process Parameters and the Formation of GTAW Additive Manufacturing of TC4 Titanium Alloy Using the Response Surface Method. *Coatings*, 1578-1593.

Manna, R. (2012). *Continuous Cooling Transformation (CCT) Diagrams*. (Banaras Hindu University) Retrieved 11 15, 2020, from https://www.phase-trans.msm.cam.ac.uk/2012/Manna/Part3.pdf.

Martukanitz, R., Michaleris, P., Palmer, T., DebRoy, T., Liu, Z.-K., Otis, R., . . . Chen, L.-Q. (2014). Toward an Integrated Computational System for Describing the Additive Manufacturing Process for Metallic Materials. *Additive Manufacturing*, 52-63.

Meyers, M., & Chawla, K. (2009). *Mechanical behavior of Materials.* Cambridge: Cambridge University Press.

Michaleris, P. (2014). Modeling Metal Deposition in Heat Transfer Analysis of Additive Manufacturing Processes. *Finite Element in Analysis and Design*, 51-60.

Mostafaei, M., & Kazeminezhad, M. (2016). Microstructural Evolution During Ultra-Rapid Annealing of Severly Deformed Low-Carbon Steel: Strain, Temperature, and Heating Rate Effects. *International Journal of Minerals, Metallurgy and Materials*, 779-793.

N. Saeidi, A. (2009). Comparison of Mechanical Properties of Martensite/Ferrite and Bainite/Ferrite Dual Phase 4340 Steels. *Materials Science and Engineering*, 125-129.

Navarro-Lopez, A., Hidalgo, J., Sietsma, J., & Santofimia, M. (2017). Characterization of bainitic/martensitic structures formed in isothermal treatments below the Mstart Temperature. *Materials Characterization*, 248-256.

Obasi, G., Pickering, E., Vasileiou, A., Sun, Y., Rathod, D., Preuss, M., . . . Smith, M. (2019). Measurement and Prediction of Phase Transformation Kinetics in a Nuclear Steel During Rapid Thermal Cycles. *Metallurgical and Materials Transactions*, 1715-1731.

Pattanayal, S., & Sahoo, S. (2021). Gas Metal Arc Welding Based Additive Manufacturing - A Review. *Journal of Manufacturing Science and Technology*, 398-442.

Rodriguez, N., Vazquez, L., Huarte, I., Arruti, E., Tabernera, I., & Alvarez, P. (2018). Wire and Arc Additive Manufacturing: A comparison Between CMT and TopTIG Processes Applied to Stainless Steel. *Welding in the World*, 1083-1096.

Schneider, J. (2020). Comparison of Microstructural Response to Heat Treatment of Inconel 718 prepared by Three Different AM Processes. *The Journal of the Minerals, Metals and Materials Society (TMS)*, 1085-1091.

Schneider, J., & Gradl, P. (2022). Directed Energy Deposition Moves Outside the Box. *Advanced Materials and Processes*, 13-18.

Shang, S., Wellburn, D., Sun, Y. Z., Wang, S. Y., Cheng, J., Liang, J., & Liu, C. S. (2014). Laser Beam Profile Moduluation for Microstructure Control In Laser Cladding of an NiCrBSi Alloy. *Surface and Coatings Technology, 248*, 46-53.

Shinha, A., Pramanik, S., & Yagati, K. (2023). Effect of Interlayer Time Interval on GTAW based Wire Arc Additive Manufacturing of 2319 Aluminum Alloy. *Sadhana*, 122-134.

Souza Neto, F., Neves, D., Silva, O., Lima, M., & Abdalla, A. (2015). An Analysis of the Mechancial Behavior of AISI 4130 Steel After TIG and Laser Welding Processes. *Procedia Engineering*, 181-188.

Stockman, T. (2019). *An Industrually Applicable Approach to Transient Thermal Modeling and Process Control in Additive Manufacturing Using a Mass-added Finate Difference Method.* PhD Disertation.

Stone, J. (2020). *Verifying Predictive Temperature Gradients for an As-Built Additively Manufactured Part.* Masters book.

Sun, S. D., Fabijanic, D., Barr, C., Liu, Q., Walker, K., Matthews, N., . . . Brandt, M. (2018). In-situ Quench and Tempering for Microstructure Control and Enhanced Mechanical Properties of Laser Cladded AISI 420 Stainless Steel Powder on 300M Steel Substrates. *Surface and Costing Technology, 333*, 210-219.

Tammas-Williams, S., & Todd, I. (2016). Design for Additive Manufacturing With Site-Specific Properties In Metlas and Alloys. *Scripta Materiala*, 105-110.

Verein, D. (1992). *Steel - A Handbook for Materials Research and Engineering.* Berlin: Springer Verlag.

Vijayan, D., & Rao, S. (2018). Process Parameter Optimization in TIG Welding of AISI 4340 Low Alloy Steel Welds by Genetic Alforithm. *Materials Science and Engineering*, 390-398.

WA. Alloy Corporation. (n.d.). *ER4340 Welding Wire and Rod Data Sheet.* WA. Alloy Co.

Wang, F., Williams, S., Colegrove, P., & Antonysamy, A. (2013). Microstructure and Mechanical Properties of Wire and Arc Additive Manufactured Ti6Al-4V. *The Minerals, Metals and Materials Society*, 968-977.

Wang, X., Wang, A., & Li, Y. (2020). Study on the Deposition Accuracy of Omni-Directional GTAW-based Additive Manufacturing. *Materials Processing Technology*.

Wang, X., Wang, A., Wang, K., & Li, Y. (2019). Process Stability for GTAW-Based Additive Manufacturing. *Rapid Prototyping Journal*, 809-819.

Williams, S. W., Martina, F., Addison, A., Ding, J., Pardal, G., & Colegrove, P. (2016). Wire + Arc Additive Manufacturing. *Material Science and Technology*, 641-647.

Wu, B., Pan, Z., Ding, D., Cuiuri, D., Li, H., & Xu, J. (2018). A Review of the Wire Arc Additive Manufacturing of Metals: Properties, Defects and Quality Improvement. *Journal of Manufacturing Processes*, 127-139.

Xu, T., Tang, S., Liu, C., Li, Z., Fan, H., & Ma, S. (2020). Obtaining Large-Size Pyramidal Lattice Cell Structures by Pulse Wire Arc Additive Manufacturing. *Materials and Design*.

Yilmaz, O., & Ugla, A. A. (2016). Shaped Metal Deposition Technique in Additive Manufacturing: A Review. *Journal of Engineering Manufacture*, 1-18.

Zappa, S., Hoyos, J., Tufara, L., & Svoboda, H. (2022). Effect of Heating Rate on Martensite to Austenite Transformation Kinetics in Supermartensitic Stainless Steel Weld Deposit. *Journal of Materials Engineering and Performance*, 8668-8676.

Zhang, H., Huang, J., Liu, C., Ma, Y., Han, Y., Xu, T., . . . Fang, H. (2020). Fabricating Pyramidal Lattice Structures of 304 L Stainless Steel by Wire Arc Additive Manufacturing. *Materials*.

Zheng, B., Zhou, Y., & Smugeresky, J. E. (2009). Thermal Behavior and Microstructural Evolution During Laster Deposition with Laser Engineered Net Shaping. *Metallurgical and Materials Transactions*, (pp. 2228-2236).

Appendix A. DATM Post Processing MATLAB Code

```matlab
close all; % Closes all open plots to ensure they are reset

clc; % Clears Command Window

testdata=1; % 1 to compare to test data

NodeQTY=4; % Number of DAS nodes created in DATM

set(0,'DefaultFigureWindowStyle','docked') % Sets the output figures to open in a single
window instead of each in a seperate window

Directory=uigetdir; % User interface to select folder containing data

File=dir(fullfile(Directory))  % Saves the file path of data folder

Length=length(File);

i=1;

for k=(Length-(NodeQTY)):Length-1 % This FOR loop imports the DATM Python Data
into a usable matlab format

    File(k,1).name;

    tempimport2=py.open(File(k,1).name,'rb');

    tempimport=py.pickle.load(tempimport2);

    temp(i,:)=cell2mat(cell(tempimport));

    i=i+1;

end

timeimport=py.open('time','rb'); % Pulls time file from DATM Python output

timeimport=py.pickle.load(timeimport);

time=cell2mat(cell(timeimport));

tempK=(temp-273.5)*(9/5)+32; % Converts the DATM temperatures in Kelvin to F
```

```matlab
set(gcf,'Visible','on'); % Pops plots out of Matlab Live Editor

set(gcf,'color','w'); % Sets the plot background color to white

for j=1:NodeQTY % Plots the thermal time history for each node on a seperate plot

    set(gcf,'Visible','on');

    figure(j)

    nodenum=string(j);

    tit1='Node';

    tit1=[tit1,nodenum]; % Creates Figure Identifier to aligne with node number

    tit1=strjoin(tit1);

    plot(time,tempK(j,:)) % Plots Data

    xlabel('Time (s)') % Labels X axis

    ylabel('Temp (F)') % Labels Y axis

    title(tit1); % Creates plot title

    set(gcf,'Visible','on');

end

figure(j+1); % Creates a new figure after individual node figures are generated

set(gcf,'color','w');

for k=1:4

    set(gcf,'color','w');

    plot(time,tempK(k,:)) % This for loop plots all node data on a single plot with
individual traces

    set(gcf,'Visible','on')

    hold on
```

    xlabel('Time (s)')

    ylabel('Temperature (F)')

end

legend('1','2','3','4','5','6','7','8') % Legend entries for nodes, extra entries are ignored

Compare prediction to test data

% Import test data matt file first as variable "A"

StartPoint=0; % This is used to set the "start" time for the test data since the model

always starts at 0 sec

RefNode = 1; % This selects which node the test data will be compared to

TCNum=2; % This selects which TC from the test data is used

samplerate=1/2; % Data sample rate 1/Hz

TempOffset=0; % This is used to align the starting temperature of the model to the test

data to account for changes in ambient temp

timeend=0; % This limits how much of the test data is used since there are long cooling

times

AF=(A*9/5)+32; % Converts recorded temp in C to F for

if testdata==1

    time2=0:samplerate:length(A)*samplerate-samplerate; % Creates time vector from

sample rate

    time2=time2-StartPoint % Adjusts start of data to 0

    figure(j)

    plot(time2,AF(:,TCNum),time,tempK(RefNode,:)-TempOffset) % Plots model data

and test data on the same plot

```matlab
xlim([0,max(time2)-timeend]) % Sets limits of x axis

ylim([0,1000]) % Sets limits of y axis

set(gcf,'Visible','on')

set(gcf,'color','w')

xlabel('Time(s)')

ylabel('Temperature (F)')

legend('Measured','Predicted')

title('Insert Plot Title') % Replace "insert plot title" with desired title of plot
end
```

Appendix B. DATM CCT Curve Generation of Model Overlay MATLAB Code

```
ModelData=1;

ModelTimeS=2428;

ModelTimeE=2648;

TempNode=3;

ModelTime=time(1,ModelTimeS:ModelTimeE);

ModelTime=ModelTime-ModelTime(1,1);

ModelTemp=tempK(TempNode,ModelTimeS:ModelTimeE);

TestData=0;

Time_Axis_Pixel=[152; 227; 303; 380; 456; 532; 617]; %pixels in x direction

Time_Axis_Time=[1; 10; 100; 1000; 10000; 100000; 1000000]; %seconds

Temp_Axis_Pixel=[683 653 610 566 523 481 436 394 351 307 265 222 179 136 92];
%pixels in y direction

Temp_Axis_Temp=[0 100 200 300 400 500 600 700 800 900 1000 1100 1200 1300
1400]; % Deg F

Line_1_Time_Pixel=[170 191 204 216 226 234 241 248 253 258 262 266 269 272 275
277 280 282 282 286 287 288 290 291 292 293 294];

Line_1_Temp_Pixel=[133 149 162 179 196 213 232 249 269 289 306 325 344 364 384
403 423 441 460 501 519 540 558 579 598 618 636];

Line_1_Time=0.0114*exp(0.0299*Line_1_Time_Pixel);

Line_1_Temp=-2.3372*Line_1_Temp_Pixel+1619;

Line_2_Time_Pixel=[255 273 292 308 324 337 347 355 361 366 369 371 373 375 377
379 381 384 386 392 394 395 397 399 401 403];
```

Line_2_Temp_Pixel=[135 141 150 159 172 186 202 219 239 258 277 297 316 335 354 375 395 433 454 511 531 549 570 589 609 629];

Line_2_Time=0.0114*exp(0.0299*Line_2_Time_Pixel);

Line_2_Temp=-2.3372*Line_2_Temp_Pixel+1619;

Line_3_Time_Pixel=[342 360 377 391 403 417 425 433 440 446 451 454 457 460 463 466 472 473 475 477 479 479 481 483 483 485 486];

Line_3_Temp_Pixel=[135 143 153 166 178 200 214 230 249 268 288 308 325 345 365 385 442 462 481 500 520 540 559 579 598 619 636];

Line_3_Time=0.0114*exp(0.0299*Line_3_Time_Pixel);

Line_3_Temp=-2.3372*Line_3_Temp_Pixel+1619;

Line_4_Time_Pixel=[390 408 424 452 463 472 480 487 492 496 500 503 506 509 511 513 516 518 520 522 523 524 526 527 528 529 531 531];

Line_4_Temp_Pixel=[135 142 153 180 198 214 232 251 271 287 308 326 347 365 385 404 424 445 463 484 502 522 541 561 580 599 619 635];

Line_4_Time=0.0114*exp(0.0299*Line_4_Time_Pixel);

Line_4_Temp=-2.3372*Line_4_Temp_Pixel+1619;

Line_5_Time_Pixel=[153 608];

Line_5_Temp_Pixel=[131 131];

Line_5_Time=0.0114*exp(0.0299*Line_5_Time_Pixel);

Line_5_Temp=-2.3372*Line_5_Temp_Pixel+1619;

Line_6_Time_Pixel=[611 591 576 563 542 527 510 487 468 451 441 425 408 395 384 369 363 358 355 353 353 358 362 366 373 379 388 397 407 416 425 431 443 454 462 468 474 475 477 479 483 488 498 508 517 527 536 547 555 567 579 591 602 612];

Line_6_Temp_Pixel=[109 110 111 112 115 116 120 125 131 137 141 146 155 163 170
183 191 199 203 208 211 214 216 217 219 221 222 224 226 227 227 228 227 226 225 223 222
212 205 197 189 185 178 174 170 165 162 159 157 154 151 148 146 144];

Line_6_Time=0.0114*exp(0.0299*Line_6_Time_Pixel);

Line_6_Temp=-2.3372*Line_6_Temp_Pixel+1619;

Line_7_Time_Pixel=[611 591 557 536 523 506 495 483 468 457 446 442 439 435 432
430 429 429 429 429 431 431];

Line_7_Temp_Pixel=[133 136 144 148 152 157 161 165 170 176 183 185 189 193 198
204 208 211 215 219 222 225];

Line_7_Time=0.0114*exp(0.0299*Line_7_Time_Pixel);

Line_7_Temp=-2.3372*Line_7_Temp_Pixel+1619;

Line_8_Time_Pixel=[273 276 279 283 290 297 304 312 319 330 341 352 362 373 400
417 436 455 474 497 611];

Line_8_Temp_Pixel=[360 352 345 339 329 321 315 310 306 302 299 295 293 292 290
289 288 287 287 287 287];

Line_8_Time=0.0114*exp(0.0299*Line_8_Time_Pixel);

Line_8_Temp=-2.3372*Line_8_Temp_Pixel+1619;

Line_9_Time_Pixel=[154 615];

Line_9_Temp_Pixel=[638 638];

Line_9_Time=0.0114*exp(0.0299*Line_9_Time_Pixel);

Line_9_Temp=-2.3372*Line_9_Temp_Pixel+1619;

Line_10_Time_Pixel=[152 518];

Line_10_Temp_Pixel=[461 461];

```matlab
Line_10_Time=0.0114*exp(0.0299*Line_10_Time_Pixel);

Line_10_Temp=-2.3372*Line_10_Temp_Pixel+1619;

figure(11)

semilogx(Line_1_Time,Line_1_Temp,Line_2_Time,Line_2_Temp,Line_3_Time,Line_3
_Temp,Line_4_Time,Line_4_Temp,Line_5_Time,Line_5_Temp,Line_6_Time,Line_6_Temp,Li
ne_7_Time,Line_7_Temp,Line_8_Time,Line_8_Temp,Line_9_Time,Line_9_Temp,Line_10_Ti
me,Line_10_Temp,ModelTime,ModelTemp);

axis([1 1000000 0 1500])

set(gcf,'color','w')

set(gca,'ytick',0:100:1500)

title('S43 Cooling Cuve - 4340 CCT Curve Overlay')

xlabel('Time  (s)')

ylabel('Temperature  (deg F)')

legend('','','','','','','','','','','Predicted')

hold on
```